HOW TO SHOOT
(Including Care and Preservation of the Rifle)

By Major Jas. A. Moss
United States Army

Introduction by Colonel William Libbey
President, National Rifle Association of America

**Fredonia Books
Amsterdam, The Netherlands**

How to Shoot:
(Including Care and Preservation of the Rifle)

by
Major Jas. A. Moss

Introduction by Colonel William Libbey

ISBN: 1-4101-0282-3

Copyright © 2003 by Fredonia Books

Reprinted from the 1917 edition

Fredonia Books
Amsterdam, The Netherlands
http://www.fredoniabooks.com

All rights reserved, including the right to reproduce this book, or portions thereof, in any form.

In order to make original editions of historical works available to scholars at an economical price, this facsimile of the original edition of 1917 is reproduced from the best available copy and has been digitally enhanced to improve legibility, but the text remains unaltered to retain historical authenticity.

INTRODUCTION

By Colonel William Libbey

President, National Rifle Association of America

There was a time when we could be considered a nation of rifle shots. Early in the Revolutionary period, certain prisoners of war were taken across the ocean to demonstrate the really dangerous foes we were, because of the wonderful skill at short ranges, which had been developed by the daily use of the rifle.

It was the faithful friend in times of peace or war. It provided the main articles of food for the household, and protected the fireside from intrusion. But like many another friend, when safety and plenty surrounded those early homes, and game became somewhat scarce, it was forgotten, and when remembered it was placed above the family hearth, and there revered for what it had done in the past.

Rifle shooting as a diversion was later on kept alive by a few devoted lovers of this king of outdoor sports. Then came an era of invention which made the rifle of today a remarkable scientific instrument. In the hands of a skillful student of conditions it can be made to do wonders,—far greater even than driving a tack at fifty yards.

All this is of course a mystery to the novice, but as he begins to realize the possibilities of the game, he becomes fascinated by the really scientific problems of ballistics. These problems are complex enough at times to puzzle even the most acute observer, but when mastered they are most satisfying in their results, both objectively and subjectively.

Since the recent revival of the American spirit of preparedness, many thousands of our younger men are turning back to the old friend of their fathers, now a very different looking gun from the long and heavy rifle of the early days. A piece which can be trusted at a thousand yards as fully as their ancestor's could be depended upon for less than a hundred yards. With this greater range and accuracy have come greater penetration, and our modern rifle has become a most formidable weapon of defense. It is well worth the trouble taken to master it. Aside from its value as a means of training it is a clean sport, as it demands of its votaries keen eyes, good judgment, steady nerves and strong muscles, and who ever heard of these except as the result of a clean and honest life?

This little volume should be an efficient aid in overcoming the technical difficulties involved. It should open the way to a new sport with an ancient and honorable lineage, and at the same time it will help to prepare our citizens in the successful use of their chief means of defense should the safety of our country ever be endangered.

We should always be, judged by our ancestry as well as by our own good sense, a race of citizen-soldiers, confident by reason of our skill at arms, that we can each contribute our full share to defend our institutions if needed. There is ultimately no greater guarantee of confidence than this trust in our arm of service, and the knowledge that we can use it to the uttermost.

When this period arrives we will be again respected as a nation of rifle shots.

William Libbey

President, National Rifle Association of America.
Princeton, N. J.,
September 20, 1916.

PREFATORY

This booklet is based on the corresponding chapters in "PRIVATES' MANUAL," by the Author, and is merely a presentation of the general method of instruction that is followed in teaching soldiers in the Regular Army how to shoot.

Then

That, unlike our forefathers, we are no longer a nation of shots, and that, if we would be prepared to defend our Country, our people must learn to shoot, is shown by the significant fact that of the 119,874 National Guardsmen on the rolls at the time of the President's call in connection with the mobilization on the Mexican border, 56,813, or about 47% had never fired a rifle, and more than 14,000 had received ratings of less than first classman. If this little book helps any considerable number of our people to learn how to shoot, it will have filled its mission.

Now

Jas. A. Moss.

Camp Gaillard, Canal Zone,
November 26. 1916.

INDEX

A

Adjustment of sights 65
Advantages of peep sight 17
Aiming 13; 35; 89
Aiming and position drills. See, "Position and aiming drills."
Appearance of objects 71

B

Battle sight 21
Bayonet fixed, firing with 76
Bearings and cams, care of 101
Bore:
  Care of 95
  How to clean 98
  How to oil 99
Brush and thong 92

C

Calling shots 73
Cams and bearings, care of 101
Canting piece 30
Care of:
  Bore 95
  Rifle 91
Chamber, care of 99
Cleaning rod 93
Clock designation of winds 63
Coaching 77
Cold, effect of 75

D

Danger signals 83
Deflection 62
Deflection and elevation correction drills 59
Designation of winds 63
Different kinds of sights 15
Disadvantages of peep sight 17
Disks 78
Dress and equipment 78

E

Effect of heat, cold, moisture, and light 75
Elevation and deflection correction drills 59
Equipment and dress 78
Estimating distance 69

F

Factors entering into shooting 12
Finding the target 77
Fine sight 20
Firing in pairs 77
Firing point 78; 81
Firing with bayonet fixed 76
Flinching 72
Fouling:
  How to remove 97
  Kinds 96
Front sight cover 65
Full sight 21

G

Gallery practice 66
Getting out of rifle all there is in it .. 12

H

Half-masting targets 84
Heat, effect of 75
Holding of piece and position of body:
  Kneeling 53
  Prone 56
  Sitting 55
  Standing 53
Hollifield Target Rod 89
Hoppe's Nitro Powder Solvent 94

I

Implements for cleaning rifle 91
Importance of shooting straight ... 11

K

Kneeling position, hold of rifle and position of body 53

L

Light, effect of 75
Light, effect of, on objects 71
Line of sight 15

M

Magazine, care of 100
Marking 82
Materials for cleaning rifle 91
Mechanism, care of 100
Metal parts, care of 100
Mirage 76
Moisture, effect of 75

N

Nitro solvent 94
Normal sight 20

O

Objects, appearance of 71
Oil, "3-In-One" 95
Oil, how to apply 95
Oiler and thong case 91
Oiling a barrel 99
Open Sight 15

P

Padding 73
Parapet 78
Peep sight 16
Pit 78
Point of aim 25
Points to be remembered:
  At all times 87
  Before firing 84
  While firing 85
Points to be remembered in caring for rifle 101

Position and aiming drills:
Importance 22
Kneeling 41
Object 31
Prone 44
Sitting down 43
Standing 33
Practice, importance of 88
Preservation of rifle 91
Prone position, holding of rifle and
 position of body 56

R

Range house 78
Range, target 78
Range officer 78
Rapid-fire exercise 39
Rapid-fire, gallery practice 67
Rests, use of 77

S

Scorebook 74
Screw driver 93
Shooting straight, value and importance of 11
Sight, care of 100
Sighting 13
Sighting drills 22; 23
Sights:
Open 15
Peep 16
Sight-setting drills 59
Signals, danger 83
Sitting position, holding of rifle and
 position of body 56

Sling, use of 47
Soda solution 94
Solvent, nitro 94
Squeezing trigger, practicing 89
Standing position, hold of rifle and
 position of body 53
Stock, care of 100
Swabbing solution 94

T

Target range 78
Targets, different kinds 80
Things to be remembered in caring
 for rifle 101
Thong and brush 92
Thong case and oiler 91
"Three-In-One" oil 95
Trajectory 12
Trigger-squeeze exercise 28
Trigger-squeeze, practicing 89

U

Use of rests 77
Use of sling 47

V

Value of shooting straight 11

W

What rifleman looks at when firing 17
Windage 62
Winds, designation of 63

Z

Zero of rifle 65

HOW TO SHOOT

Chapter I

HOW TO SHOOT[1]

Value and importance of shooting straight. The value of a soldier as a fighting man is measured by his ability to shoot straight. If you can't shoot, you have no business on the firing line,—you merely take up room without accomplishing anything. In other words, you are in the way, and would be better off at home where you would not be risking your life unnecessarily, and would not be a useless burden to your officers.

From every standpoint it is to your advantage to learn to shoot. Not only does it mean more pay to you,[2] but it may some day mean more to you than all the riches in the world,—*it may save your life.*

If you ever go into battle, the consciousness that you can shoot as well as the other fellow, if not even a little better, will give you a comforting feeling of confidence that will mean more to you than all the extra pay you may have gotten for qualification in marksmanship. This comforting feeling of confidence will repay you a thousand times over for all the time, care and patience that you may have devoted to making yourself a good shot.

Remember, although you may not actually hit the other fellow, if you can shoot straight enough to make your bullets pass close to him, you will make him so

---

[1] The author is indebted to Capt. John W. Lang, 29th Inf., for valuable assistance in the preparation of this chapter.

[2] If you qualify as marksman, you get $2 a month extra; as sharpshooter, $3; as expert rifleman, $5.

nervous that he will drop down behind cover and quit shooting, or his shots will all go wild.

Any man of normal eyesight and fair intelligence can, with determination and proper instruction, become a fair shot if not an excellent one.

Factors that enter into shooting. While there are a number of factors, some very important, others less so, that enter into shooting (for example, effect of light and wind, hold of piece, trigger squeeze, physical condition, etc.), none of them is especially difficult, and they can all be mastered by *determination and practice*.

Getting out of the rifle all there is in it. Our rifle is the best and most accurate rifle in the world.

There are certain things that it is capable of doing, *that it can be made to do*.

Whether the soldier can make his rifle do what it is capable of doing,—whether he can get out of it everything there is in it (all the hidden 4's and 5's),—whether he can make it come pretty nearly doing what he wants it to do,—depends upon the soldier's *determination and instruction*.

In other words, with his officers and noncommissioned officers to instruct him, it is entirely up to the soldier himself as to whether or not he becomes at least a marksman.

The trajectory. As the bullet passes through the air it makes a curved line something like this:

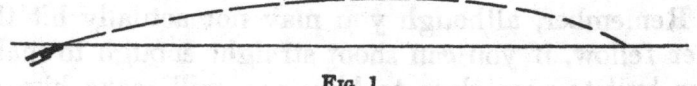

Fig. 1

This curved line is called the *trajectory*.

How to Shoot

The resistance of the air and the force of gravity (the force that pulls all bodies toward the earth), are the two things that make the path of the bullet a curved line, just the same as they make the path of the baseball thrown by the player a curved line.

The resistance of the air holds the bullet back and the force of gravity pulls it down, so that the two acting together make the bullet's path curved.

The longer the range the more will the path of the bullet (the trajectory) be curved, as shown by the following drawing:

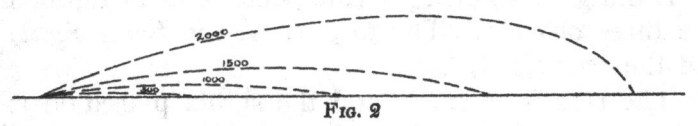

FIG. 2

The principle involved is the same as that involved in throwing a baseball. For example, if you throw a baseball very hard from third to first base, you can make it reach first base in almost a straight line, without going very high in the air, but if you wanted to throw the ball home from the outfield, you would have to throw it pretty high in order to get it there and its path (trajectory) would be curved very much. In other words, you've got to make allowance for the resistance of the air and the force of gravity.

An expert ball player knows, through practice, just how high it is necessary to throw a ball in order for it to reach certain points. A beginner does not know.

Sighting or Aiming. Now, on the rifle there are two "sights,"—the *front sight* and the *rear sight*,—which enable the rifleman to regulate the path of the bullet, as the ball player regulates the path of the ball.

If the ball player wants distance, he throws the ball high (raises the path, the trajectory), using his eye and guesswork, and likewise if the rifleman wants to shoot at a distant target, he, too, shoots the bullet high (that is, he raises the muzzle of his rifle), but he doesn't have to depend upon guesswork. It is all worked out for him by experts and all he need do is to set the *rear sight* for the proper range,—that is, for the distance the object is from him.

Aiming or sighting a rifle consists in bringing into line three objects: The *target,* A, the *front sight,* B, and the *rear sight,* C.

The rifle is so made and the sights placed on it in such a way that when the piece is held in such a position that the *target,* the *front sight* and the *rear sight* are in line, and the trigger is pulled (squeezed) the bullet will strike the *target.*

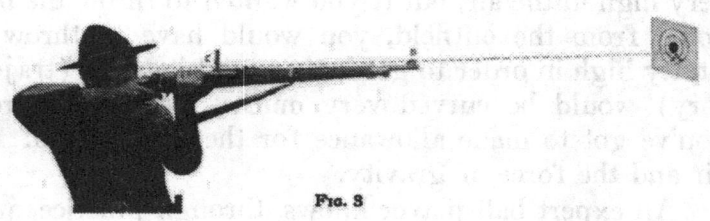

Fig. 3

You raise the muzzle of the piece by raising the rear sight,—that is, raising the rear sight has the effect of raising the muzzle, for the higher you raise the rear sight the higher must you raise the muzzle in order to see the front sight and get it in line with the object aimed at and the rear sight.

This is shown in the following illustrations:

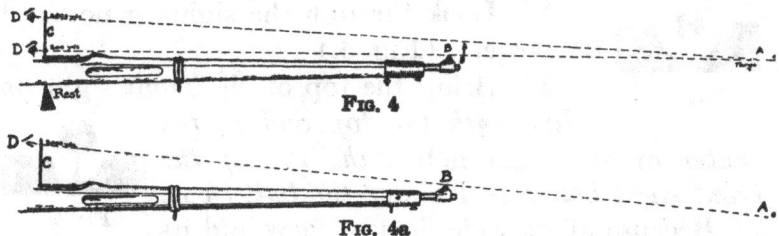

The rear sight, C, the front sight, B, and the bull's eye, A, are all on a line with the eye, D, the rear sight being set for 200 yards.

Suppose we wanted to shoot at 2000 instead of 200 yards. We would raise the slide up to 20 (2000 yards) on the sight leaf.

In order to see the bull's eye through the notch sight at 2000, we must raise the eye to the position, D. We now have the rear sight, the bull's eye and the eye in line, but we must bring the front sight in line with them, which is done by raising the muzzle of the piece, giving the result shown in Fig. 4a.

Line of sight. With the open sight the line of sight is determined by a point on the middle line of the notch of the rear sight and the top of the front sight.

With the peep sight, the line of sight is determined by the *center* of the peep and the top of the front sight.

Different kinds of sights. The different kinds of sights are as follows:

Open sight. By *open sight* is meant the use of any one of the *sighting notches*.

How to Shoot

To use the open sight:

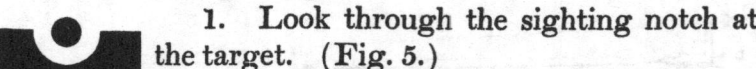
Fig. 5

1. Look through the sighting notch at the target. (Fig. 5.)
2. Bring the top of the front sight *on a line with the top and in the center* of the sight notch, *the top of the front sight being just under the bull's eye.*

Fig. 6

Because of its wide field of view and its readiness in getting a quick aim with it, the open sight is the one that is generally used in the later stages of battle, or when fire is to start immediately.

The following positions of the front sight are *incorrect*:

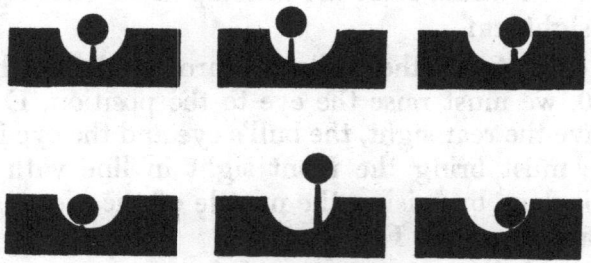
Fig. 6a

Peep sight. By *peep sight* is meant the use of the *peep hole* in the drift slide.

To use the *peep sight:*
1. Look through the peep hole at the target. (Fig. 7.)
2. Bring the top of the front sight to the *center* of the peep hole, *the top of the front sight being just under the bull's eye.* (Fig. 8.)

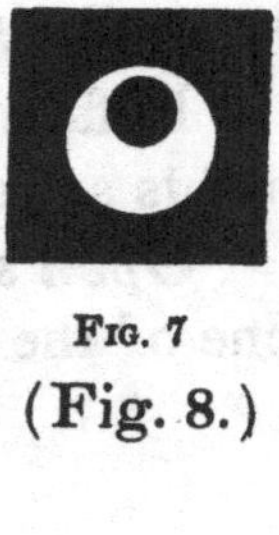

Fig. 7

Be sure to get the top of front sight, as in Fig. 8, *and not the bull's eye,* as in Fig. 9, *in center of the peep hole.*

Fig. 8
Correct

Fig. 9
Incorrect

Advantage of the peep sight. The advantage of the peep sight over the open sight is due to the fact that it is easier to center the top of the front sight in the peep hole and thus get the same amount of front sight each time.

For example, you know at once, without measuring, that the dots in the circles, Fig. 10, are not cen-

Fig. 10

tered, and that the one in the circle in Fig. 11, is. After a little practice, in looking through the peep hole the eye almost automatically centers the top of the front sight.

Fig. 11

Disadvantage of the peep sight. The disadvantage of the peep sight is that its limited field of view and lack of readiness in getting a quick aim with it limit its use to those stages of the combat when comparative deliberation will be possible.

What the rifleman looks at when he fires. The eye can be focused accurately upon objects at only one distance at a time; all other objects we see will be more or less blurred and fuzzy looking, depending upon their distance from the object upon which our eye is focused. Hold your finger up and look at it. You will see your finger clearly,—the rest of the hand and the arm will be more or less blurred,—objects about you will be seen only indistinctly. Hold your hand in place, but focus your eye on some object beyond your finger on a line with your finger and your eye. You will still see your finger but it will now be fuzzy and indistinct.

In shooting we have three points which are placed in a line—the rear sight, the front sight and the target. It is impossible to focus the eye on all three at the same time. One must be chosen.

Which shall we choose?

The following illustrations show the appearance of the bull's eye, depending upon whether the eye is focused on the front sight, rear sight or bull's eye.

In Fig. 12 the eye is focused on the *bull's eye*. Notice how clear cut and distinct it is, and the blurring of the front and rear sights.

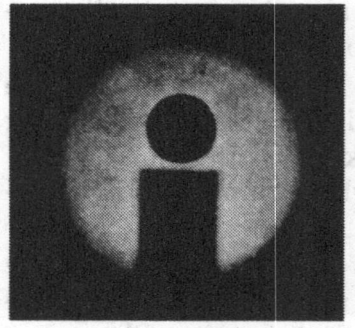

Peep normal sight   *Open normal sight*

Fig. 12

In Fig. 13 the eye is focused on the *front sight*. Notice how clear cut and distinct it is, and the blurring of the bull's eye.

 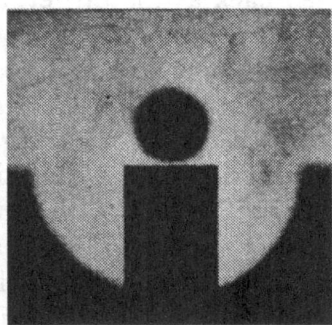

Fig. 13

In Fig. 14 the eye is focused on the *rear sight*. Notice how clear cut and distinct it is, and how blurred the front sight and the bull's eye are.

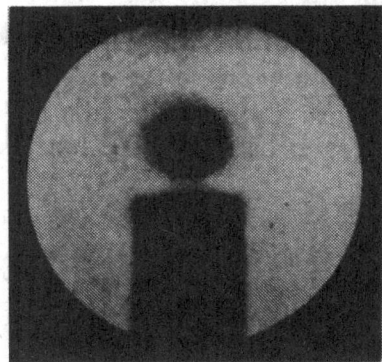 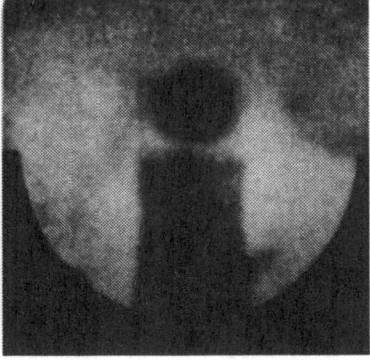

Fig. 14

The rifleman who attains proficiency *focuses his eye on the target while aiming,* but he glances at one sight and then the other to see that they are aligned properly, then back at the target, and at the instant of discharge *his eye is on the target.*

Blurring is best overcome by using the peep sight, which may be compared to looking through a round window,—whatever blurring there is will be uniform and concentric and we can still center the TOP OF THE FRONT SIGHT without difficulty.

Normal sight. The amount of front sight taken in Figs. 12, 13 and 14, is called the normal sight and is the one that the soldier should always use, either with the open notch or peep sight, as it is the only sight which assures the taking of the same amount of front sight every time. In other words, it assures a greater degree of *uniformity* in sighting, which is one of the most important factors in shooting. By uniformity in sighting is meant taking the same amount of sight each time.

If you take *less* than the amount of front sight used in the normal sight, it will, of course, have the effect of *lowering* the muzzle of the piece, and consequently you will hit a point *lower* than if you had used the normal sight.

On the other hand, if you take *more* than the amount of front sight used in the normal sight, it will, of course, have the effect of *raising* the muzzle and consequently you will hit a point *higher* than if you had used the normal sight.

Fine sight. Although occasionally a man will be found who can get good results by using the fine sight, the average man cannot, and this form of sighting is, therefore, to be avoided.

FIG. 15
Fine sight

How to Shoot

Full sight. The so-called *full sight* must be avoided under all circumstances. It is merely mentioned and shown here to point out a fault that must be carefully avoided.

The objections to its use are the same as in the case of the fine sight,—that is, lack of uniformity in the amount of sight taken.

Fig. 16
Full sight

Battle sight. By *battle sight* we mean the position of the rear sight with the leaf down. There is a sighting notch on the top of the leaf, or rather on top of the leaf slide which works up and down the leaf.

The battle sight is the only sight used in *rapid fire*. In unexpected, close encounters the side that first opens a rapid and accurate fire has a great advantage over the other. Again, a soldier on patrol generally has no time to set his sight, if suddenly attacked at close range. The battle sight, may, therefore, be called the *emergency sight,—*the *handy, quick sight.* The soldier should, therefore, become thoroughly familiar with the use of this sight.

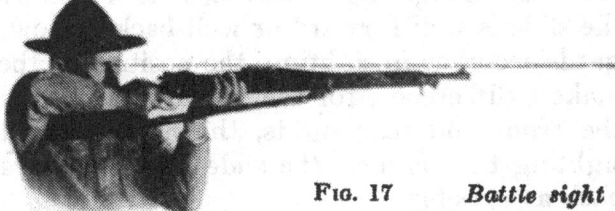

Fig. 17 *Battle sight*

The *sighting notch* in the slide with the rear sight leaf down, is the same height as is the sighting in the drift slide when the rear sight leaf is raised and set at 530 yards.

That is to say, *battle sight* is equivalent to a sight setting of 530 yards. Therefore, in shooting with battle sight at objects nearer than 530 yards you must aim lower.

The following shows the trajectory of the bullet when *battle sight* is used:

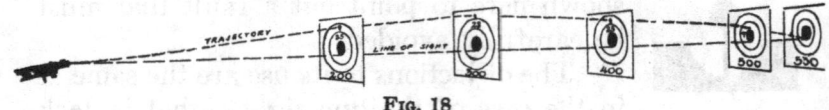

Fig. 18

That is to say, if you were aiming with the battle sight at an object 530 yards away, the bullet would pass 25 inches above an object of the same height at 200 yards, 28 inches above at 300, 23 inches above at 400, and 7 inches above one at 500, which is only another way of saying if shooting with battle sight at an object 200 yards away, you must aim 25 inches (about 2 feet) below the object in order to hit it; if at 300 yards, 28 inches (2½ feet below); if at 400 yards, 23 inches (about 2 feet) below; and if at 500 yards, 7 inches (about ½ foot) below.

Remember that in the case of the battle sight, the position of the slide is immaterial, except as regards accuracy in sighting,—that is, it is immaterial whether the slide is well forward or well back. However, as regards accuracy in sighting, the position of the slide does make a difference, for the greater the distance between the front and rear sights, the more accurate will the sighting be. Hence, the slide should always be as far back as possible.

SIGHTING, POSITION AND AIMING DRILLS

The importance of the following sighting, position and aiming drills cannot be overestimated. If they are carefully practiced, before firing a single shot at a tar-

How to Shoot

get, you will have learned how to aim your piece correctly, hold your rifle steadily, squeeze the trigger properly, assume that position best adapted to the particular conformation of your body, and you will also have acquired the quickness and manual skill required for handling the piece in rapid fire.

The sighting, position and aiming drills teach the fundamental principles of shooting, which are the foundation upon which marksmanship is built.

Do not confine yourself to going through these drills only during drill hours, but go through them frequently at other times. The extent to which it will improve your shooting will more than repay you for your trouble.

Sighting Drills

Object. The objects of the sighting drill are:

1. To show how to bring the rear sight, the front sight and the target into the same line,—that is, to show how to sight properly.

2. To discover and point out errors in sighting,—in other words, to discover the errors you make in sighting and show the reasons for same, so that you may be able to correct them properly.

3. To teach uniformity in sighting,—that is, to teach you how to take the same amount of sight each time,—to see every time the same amount of front sight when you look through the rear sight.

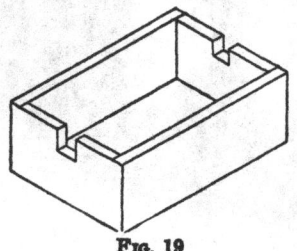
Fig. 19

Sighting rest for rifle. A good sighting rest for a rifle may be made by removing the top from an empty pistol ammunition box, or a similar box, and then cutting notches in the ends of the box to fit the rifle closely. (Fig. 19.)

Place the rifle in these notches with the trigger guard close to and outside one end.

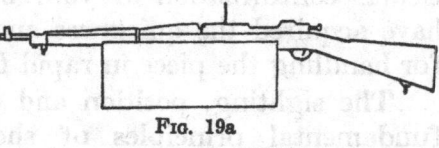

FIG. 19a

At a convenient distance above the ground fasten a blank sheet of paper on a wall or on a plank nailed to a stake driven into the ground. Three legs are fastened to the rest (or it may be placed on the ground without any legs), which is placed 20 or 30 feet from the blank sheet of paper.

FIG. 20

How to Shoot

Make sure that the piece is canted neither to the right nor left, and without touching the rifle or rest, sight the rifle near the center of the blank sheet of paper. (Fig. 20.) Changes in the line of sight are made by changing the elevation and windage.

A soldier acting as marker is provided with a pencil and a small rod bearing at one end a small piece of white cardboard, with a black bull's-eye, pierced in the center with a hole just large enough to admit the point of a lead pencil.

FIG. 21

The soldier sighting directs the marker to move the disk to the right, left, higher, or lower, until the line of aim is established when he commands, "Mark," or "Hold."

At the command "Mark," being careful not to move the disk, the marker records through the hole in the center the position of the disk and then withdraws it.

At the command "Hold," the marker holds the disk carefully in place without marking, until the position is verified by the instructor, and the disk is not withdrawn until so directed.

Point of Aim. Always be sure to aim at a point just below the black bull's-eye,—that is, aim so that there will be a fine line of light between the bottom of the bull's-eye and the top of the front sight (Fig. 22). This is important to insure uniformity in sighting,—that is, in order to make sure that the same amount of the front sight is taken each time. If the top of the front sight touches the bottom of the bull's-eye it is impossible to say just how much of the front sight is seen.

FIG. 22

First Sighting Exercise

Using the sighting rest for the rifle (Fig. 20, page 24) require each man to direct the marker to move the disk until the rifle is directed on the bull's-eye with the *normal* sight and command, "Hold." If aiming correctly the rear sight, the front sight and the bull's-eye will look as shown in Fig. 22, above.

The instructor then verifies this line of sight. Errors, if any, will be pointed out to the soldier and another trial made. If he is still unable to sight correctly, he will be given as many more trials as may be necessary.

Sometimes a man does not know how to place the eye in the line of sight; he will look over or along one side of the notch of the rear sight and believe that he is aiming through the notch because he sees it at the same time that he does the front sight. Again some men in sighting will look at the front sight and not at the object.

FIG. 23

Repeat the above exercise, using the *peep* sight. If aiming correctly, the rear

How to Shoot

sight, the front sight and the bull's-eye will look as shown in Fig. 23.

Second Sighting Exercise

The triangle of sighting. Using the sighting rest for the rifle as before (Fig. 20, page 24), direct the marker to move the disk until the rifle is directed on the bull's-eye with the *normal* sight and command "Mark," whereupon the marker, being careful not to move the disk, records through the hole in its center, the position of the disk, and withdraws it. Then, being careful not to move the rifle or sights repeat the operation until three marks have been made.

Join the three points by straight lines. The shape and size of the triangle will indicate the nature of the variations made in sighting.

For example, if you have taken the same aim each time, you will get a very small triangle something like this: which resulted from taking each time this aim, for instance:

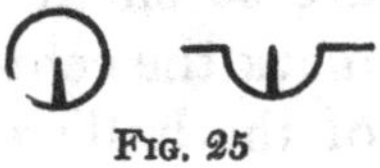

FIG. 25

A triangle like Fig. 26 results from not taking the same amount of front sight each time, as shown in Fig. 27.

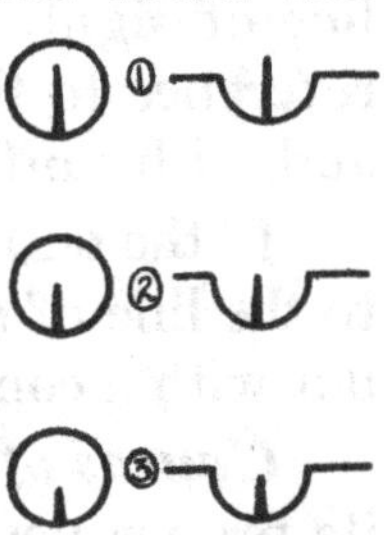

FIG. 26

A triangle like Fig. 28 shows that the front sight was not in the middle of the notch each time, as shown in Fig. 29.

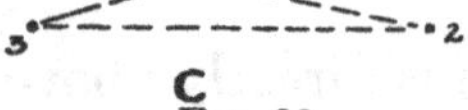

FIG. 28

FIG. 27

A triangle like Fig. 30 results from a combination of the two errors mentioned above,—that is, not taking the same amount of front sight each time and not having the front sight in the middle of the notch each time, as shown in Fig. 31.

Fig. 30

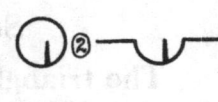

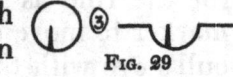
Fig. 29

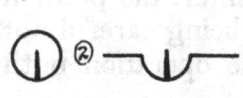

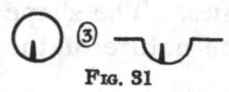
Fig. 31

If any one of the sides of the triangle is longer than one-half inch, the exercise is repeated, each sight being verified by the instructor, who will call the soldier's attention to his errors, if any.

The smaller the triangle, the better the sighting.

Verifying the triangle. If the sides of the triangle are so small that they indicate regularity in sighting, mark the center of the triangle and then place the center of the bull's-eye on this mark. The instructor then examines the position of the bull's-eye with reference to the line of sight. If the bull's-eye is properly placed with reference to the line of sight, the soldier aims correctly and with uniformity.

If the bull's-eye is not properly placed with reference to the line of sight, the soldier aims in a regular manner but with a constant error.

Causes of errors. If the bull's-eye is directly above its proper position, the soldier has aimed high,—that is, he has taken too little front sight.

How to Shoot

If the bull's-eye is directly below its proper position, the soldier has aimed low,—that is, he has taken too much front sight.

If the bull's-eye is directly to the right or left of its proper position, the soldier has not sighted through the center of the rear notch and over the top of the front sight. If to the *right,* the soldier has either sighted along the *left* of the rear sight notch or the *right* side of the front sight, or has committed both of these errors.

If the bull's-eye is to the *left* of its proper place, the soldier has probably sighted along the *right* of the rear sight notch, or to the *left* of the front sight, or has committed both of these errors.

If the bull's eye is diagonally above and to the right, the soldier has probably combined the errors which placed it too high and too far to the right.

Any other diagonal position would be produced by a similar combination of vertical and horizontal errors.

After the above instruction has been given to one man, the line of sight will be slightly changed by moving the sighting rest or by changing the elevation and windage, and the exercises similarly repeated with other men.

Repeat the exercise, using the *peep* sight.

Third Sighting Exercise

This exercise shows the effect of canting the piece.

It is most important that in aiming the sights be kept vertical and the piece not be canted,—that is, that the barrel be not tilted over to the right or left.

If the piece is canted to the right, the sights are lowered to the right and consequently the bullet will strike to the right and below the point aimed at, even though the rifle be otherwise correctly aimed and the sights correctly set.

Similarly if the piece is canted to the left the sights are lowered to the left, and consequently the bullet will strike to the left and low.

This effect of canting the piece may be shown as follows: Use the sighting rest with the rifle firmly held in the notches, the bolt removed.

Paste a black paster near the center of the bottom line of the target. Sight the rifle on this mark, using about 2000 yards' elevation. Then, being careful not to move the rifle, look through the bore and direct the marker to move the disk until the bull's-eye is in the center of the field of view and command, "Mark."

Next, turn the rest (with the rifle) over 90° to the right, on its side, and with the same elevation, sight on the same paster as above. Then, being careful not to move the rifle, look through the bore and again direct the marker to move the disk until the bull's-eye is in the center of the field of view and command, "Mark."

Not considering the fall of the bullet, the first mark represents the point struck with the sight vertical, the second mark represents the point struck, low and to the right, using the same elevation and the same point of aim, when the piece is canted 90° to the right.

Different degrees of canting the piece can be represented by drawing an arc of a circle through the two marks with the paster as a center. The second mark will be at a point on this arc corresponding to the degree of canting the piece.

It is important to know that this effect of canting increases with the distance from the target.

intensified when combined with a large amount of paranoia.

There is another group of doers, the misfits of society, the indigent alcoholics and drug addicts who formerly were arrested and jailed. The laws no longer permit this. Doctors and psychiatrists don't know what to do with them and they don't know what to do with themselves, although group therapy is of help to some. They are an unhappy lot. These are not criminals, but a drain on the economy.

Still another is the type of personality that has no conscience and is totally immoral. We have become familiar with such human entities through the press and occasionally personally. How can a person go into a convenience store, hold up the proprietor, empty the cash register, then shoot him dead? How can the rapist beat up a woman, have his way with her, then slash her throat or shoot her? Why is it that these criminals are apprehended, tried and found guilty, punished with prison but then released on parole after serving a minimum term? Often they resume their lifestyles and kill, rape or rob again. And there are the serial killers, the "Son of Sam, the "Hillside Strangler and Charles Manson who murdered eight people all for no reason. Such episodes of reprehensible and contemptible murder appear in the press periodically. What motivates these people? Is there a malfunction in their brains that causes this behavior?

We call them psychopaths or sociopaths. The word is not used very often in ordinary conversation. One seldom hears it, however, it is the term for the hardened individual described above who lives in crime as his career, (the career criminal). These people are utterly incapable of experiencing the usual emotional sensations we feel, such as compassion, love, understanding, pity, guilt, remorse.

The career criminal has the "gift of gab" and a remarkable facility for talking his way out of scrapes and ingratiating himself to his listener. He uses words of love and endearment with ease, and vents emotions and sympathy but has no capacity of feeling what he is saying. He understands the words and what they mean at the intellectual level but has never and wiil never be conscious of them. If he is caught in a lie, he quickly changes the subject and shifts into another.

Psychopaths are recognized easily by applying a formula worked out by scientists who have studied them for many years. They all have the same characteristics. Aside from the traits mentioned above, the psychopath is charming and ingratiating. He introduces interesting subjects into the conversation with glibness, and an eloquence that demonstrates more than average intelligence. He can talk himself out of any situation and can fool all but the interrogator who is informed and does not fall for his line. But his words do not match his action's. His deeds are as damaging as his words are disarming.

The psychopath is also found in other fields of endeavor. The professions and business are not exempt from their share of shysters, insider traders, unethical doctors, and dishonest business men. These are evil people who prey upon society and cause inestimable hardship to innocent people in every society and in every realm of society.

Studies by scientists for which special tests were devised, show that the psychopath, eighty percent of whom are men, clearly indicate that they have no understanding nor do they experience feelings of emotion or the pangs of conscience, as we do. When they kill, they feel nothing but a sense of accomplishment, a job well done. And as well as being devoid of compassion they are also devoid of fear. Further research clearly indicates that the psychopath has a defect in the brain that points to a failure in its development.

The psychopath is a con man. He is irresponsible and manipulative, a fluent and smooth talker, naive yet selfish, charming yet callous and an easy and convincing liar. He loves to break the laws laid down by society. He is completely amoral. He lives only in the present and what it can bring him, and has no desire to plan ahead. This shows a state of immaturity which, with age usually improves. Consequently, it is the opinion of some researchers that there is hope that the psychopath can be rehabilitated through a well designed program of psychotherapy to accelerate his maturation. Up until now, it was believed by all studying the psychopath that there was no hope for treatment and he is destined to be incarcerated for the rest of his life.

At this point in history, the urgency to find the answers to criminal behavior has never been greater. Statistics show that the incidence of psychopaths in the general population is rising steadily, especially within inner cities. Their numbers now are from two to three percent of the population.

Furthermore, the crimes are being committed by younger and more brutal individuals. It is well documented that the criminal begins his escapades at home very early in life, and grows up with a history of being in and out of trouble with the law before he is an adult.

What has not been documented for lack of sufficient data is the incidence of individuals in the everyday stream of business and the professions who disrupt and destroy the well-being, sometimes even the lives, of countless of innocent people. Scientists believe that their lack of guilt, empathy and consideration is as great as that of the criminal who rapes, tortures or kills his victim. And not many of these so-called white collar criminals end up in prison. They remain moving about freely in society wreaking havoc, doing harm and profiting thereby, with rare retribution. The press is full these days of examples.

This brings us to the matter of crime and drugs. Does a criminal on drugs have an impact on the crime he commits? Is someone on drugs who is not a criminal apt to commit a crime? The experts say, yes. Although drugs have always been available to the few who turn to them, the past fifteen years show an insurgence of their use unparalelled in history. We submit that the advent of the form of cocaine called "crack" is the cause of this tremendous increase in crime. The public does not realize what crack does to the workings of the brain. Most users would probably abandon it, if they had the ability, but this addiction is usually permanent. Although drugs have been known throughout history and used by the few, young people began their social acceptance in the 1960s. Twenty years later, America became the greatest consumer of illicit drugs in the world, especially the most dangerous, cocaine.

The use of cocaine is harmful to the body as well as to the brain. Sudden death from cocaine is not uncommon and can be caused by other "hard" drugs as well, such as heroin. Cocaine is known to damage the heart, causing rupture of heart muscle, it decreases the supply of blood and oxygen to the heart, causing damage to blood vessels in the brain resulting in death. It damages the liver, kidneys and depletes the immune system, laying the body open to disease.

When cocaine is smoked, sniffed or injected it goes immediately to the brain and gives a euphoric feeling. It produces that feeling by traveling straight to the primitive Old Brain, called the "limbic" system which controls drives and emotions and the essential processes for survival, such as nutrition and reproduction. The Old Brain is surrounded by the New Brain, or "neocortex", which is the center of intelligence, reasoning powers and self consciousness. These two brains complement one another and each is necessary to the other.

Cocaine causes the release of an excessive amount of neurotransmitter substances in the brain, namely, norepinephrine and dopamine which over- stimulates and disrupts normal function. It produces compulsive and irrational behavior, and extreme paranoia. The intellect is affected, and the ability to reason, thus removing inhibitions and allowing a free rein to emotions and passions, including anger and violence.

Cocaine interferes with the body's homeostasis, lowering the sense of physical well-being and emotional stability. As yet there is no sure cure for drug abuse of any kind, although much work is being done to find it. Some people are able to drink alcohol moderately and never become alcoholics. Some take an occasional snort of cocaine or marijuana and not become addicted, but until we find the secret of addiction in the individual, it is best never to experiment with drugs, especially hard drugs. Why? Some of us are born addictive to drugs, to food, to gambling, to work or to whatever turns us on and, until we find the answer as to why and how to harness this craving, be careful.

PARANOIA, OUR PROTECTIVE TRAIT

Paranoia is the characteristic of personality that has good points and bad ones. It protects us, warming and enfolding us, and makes us courageous yet careful, self-critical yet self-assured, and cautiously defensive. It also causes us to be gossipy, deceitful and manipulating, and brings out the fearful, apprehensive, suspicious, critical and blaming elements of our characters. This paranoid quality is common to every living thing, from the tiniest insect, even plants and trees, to complex human beings. It is the instinct that protects us, acting like a sixth sense. It is all pervasive and always present, coloring everything we do and every thought we have.

So paranoia has many facets. Being fearful of physical harm is one. Another is the feeling that makes us want to control the situation and the people around us and to manipulate them in order to get control. It is also being wrapped up in ourselves and ignoring the needs of others. In essence, it is our selfish quality and influences everything we do. It is not difficult to imagine what a person with a great deal of paranoia would be like.

Every living thing is born with a primitive reaction to fear called the "fight or flight" syndrome. Paranoia arouses this, and when caught in a crisis, we have to quickly decide whether to run away or stay and fight. Wild animals could not survive without it and people and animals alike take the offensive when threatened. So paranoia is our constant companion, ready to be mobilized when needed, especially when stressed.

Some people have too little and are vulnerable to emotional threats, exploitation and fraud. They can also be childishly unafraid, and physically hurt by taking foolish risks. Last winter, the high school hockey team went to practice on the pond. They tested the ice and found it unsafe. One boy defied his friends and fell in. He has too little paranoia.

I have two cats, one has a fair amount of paranoia and the other not enough. When a stranger comes, Gappy disappears into the cellar, and Pywacket nonchalantly marches into the room and sniffs at my guest's legs. I was born with little paranoia. I was always taking chances and trusted everybody. Learning about personality helps me develop wisdom in dealing with life.

Too much paranoia makes us fearful. We are constantly afraid we will be attacked, either verbally or physically. We might attack first without provocation, because our defense mechanism takes precedence over better judgment.We distrust others without reason, and blame them for what goes wrong. We suspect people without cause and question the kindest motives. High paranoia is also linked to unrealistic anxiety and insecurity, and if a person has high energy and a doing personality as well, he can be dangerous.

The world is controlled by strongly paranoid people. Politicians and religious leaders have a lot; they have to in order to function. And scattered among the population, in public and private life, is the paranoid personality which it is best to steer clear of. Most people like this are socially acceptable despite their sarcastic leanings. This is their defense, however, and they can be hostile if called to task and have a tendency to blame others for their mistakes and shortcomings. As they grow older, this becomes more emphasized. They can be successful achievers, but they seldom find happiness, especially in close relationships because of their controlling nature.

Paranoia shows up at work, and it is not a comfortable feeling. Is your colleague criticizing you? Are people talking about you? How are you doing? Is "big brother" watching you? You are fearful, suspicious, apprehensive, selfcentered, and challenge criticism. You can even be belligerent, depending on your degree of paranoia. You may have all these reactions or none, and may feel relaxed most of the time, paranoia showing up only when you are attacked.

Jane is like this. She is a hard worker, conscientious, compulsive, even slightly fanatical. Her boss thinks the world of her, but she feels insecure and is afraid people will turn against her. She is apt to interpret a disapproving remark as being directed at her. Once, we were at a cocktail party, and she noticed two guests whispering. She thought they were talking about her. They were not.

If the thinking person has a lot of paranoia, he is defensive when dealing with others, and easily inflamed and emotional when disagreed with. Although his opponent may be wrong, he doesn't attack back on these grounds but on a personal level. As a thinking person, he is highly sensitive and his paranoia doesn't let him take criticism, so he strikes back with counter criticism or sarcasm. Thus, he keeps his self-image at a comfortable level. Sensitive people usually have a low opinion of themselves, and are apt to criticize those who don't meet their standards or expectations. This gives a false sense of superiority and is an ego booster.

The paranoid doer is compelled to exert control over others. He is not easy to live or work with as he does not like to be told what to do. *He* wants to do the telling. Nancy is that way. She is very health conscious and has studied for years how to keep well. She practices what she preaches and exercises madly, eats properly, keeps her weight down. She does not believe in drugs of any kind, so she doesn't smoke, drink alcohol or coffee. Now she is getting on in years, she continues to be healthy and fit and her paranoia has become accentuated. The problem is that she thinks everyone should do as she does and tells them so.

Such people often choose a career that allows them to exert influence over others. Religion and politics are two fields that attract paranoid doers. Extreme cases can be great accomplishers or selfish manipulators, unscrupulous and greedy, egocentric and demanding. They can be asocial, sometimes amoral and are driven to seek positions of power in which they thrive, some to do good, some to do evil. Several examples come to mind, in big business, evangelism and politics, and currently are in the public eye. Of those who have passed on, Winston Churchill was a paranoid doer for good, and Adolf Hitler for evil. It was fortunate for the world that Churchill lived at the same time as Hitler.

There are many combinations of thinking and doing working as a team. The balance of the two regulates how we feel and what we accomplish, and shifts from one to the other like a seesaw, according to what we are doing. For instance, I am at the typewriter moving my hands, but my thinking is hard at work deciding what to write, about eighty percent thinking and twenty doing. If I don't type or say what I am thinking no one would know my thoughts. So the doing bridges the gap between thinking and telling about it. In other words, what we think is never known unless we speak or write it.

Consider the difference between the two. The more thinking person is honest, dedicated, sensitive. He likes to use his brain and thrives on detail. The more doing person likes to be with people and enjoys the give and take, the acting and reacting with them. The contrast is roughly that of the salesman and the college professor. Both make use of others, but the thinker is more obvious as he manipulates them. The doer is more effective and less obvious, while paranoia is in control of both.

How much do we use of thinking and of doing? The amount of each we use at any one time fluctuates and depends on our mood and physical well-being. However, we can in an emergency or when a supreme effort is required, summon up one hundred percent of one or the other. Otherwise there is an intermix of the two elements, resulting in a personality appearance of the combined effects of each input.

For example, a champion figure skater is defending his title, and after months of practice, has the routine pretty well etched into his thinking which has become transmitted to his doing, the execution of the routine. During the performance, he is using eighty-five percent of his doing and fifteen of his thinking. If he makes even a slight mistake, the thinking increases, correcting the doing error but possibly reducing subsequent accuracy. As he continues skating, both thinking and doing quickly revert to the ratio with which he started.

When a tennis player reaches the finals at Wimbledon, he uses the same ratio of thinking and doing as the skater. This applies to the professional bank robber as well, who uses more thinking while he plans a coup, but when it comes to the robbery, the thinking gives way to the doing.

We don't encounter situations every day that require an extraordinary output of energy such as the above, or matching wits with an astute business man or performing critical surgery. However, the potential for the requirement may appear without warning or be planned in advance. The rest of the time we go about our lives at the pace we have set.

So one hundred percent of thinking and doing combined are the top of everyone's capability, the more thinking person using more thinking most of the time and the more doing person more doing. Both can call on a varied mix of the two but this takes an extra burst of energy, motivated and controlled by the brain. When the crisis is over, the balance of thinking and doing returns to its original level. Thinking and doing fall below their best potential as a result of ill health or an event like the death of a dear one or when depressed.

Thus, the balance between thinking and doing changes according to what is being done. Working at a desk requires more thinking, an athlete more doing. Each activity has a different balance but the original balance remains unchanged for life. Thus the doer is usually a better athlete and the thinker more intellectually capable. Great ability in both is rare.

What we make of life is up to us. We are born with potential in several areas and it is up to us to develop our talents and perfect our skills. What we accomplish is how we manage our lives. There are great athletes who never make it, talented actors who drop out after a success or two. There are brilliant people who get Ph.D.s and end up painting houses. There are aspiring writers who are always going to write that novel, but fritter away their lives on trivia. Why this waste of talent? It is usually a combination of factors, the unwillingness to work hard, the lack of inspiration, an extreme doing personality that doesn't give a damn, an extreme thinker lost in fantasy, the abuse of the body by alcohol or drugs, the hedonist who indulges himself, very low energy, the loss of a loved one. All block the development of talent and stop the natural flow of a person's capabilities. The least of the culprits is illness or physical handicap.

What we do may be different from what we plan, so we don't always accomplish what we think we will. Artistic creations are often unrealistic, so it is extremely difficult to carry them out, and they almost never materialize as designed. This applies to any work of art, painting, sculpture, music, literature. We know what we want to produce, but are frustrated, and even if proclaimed a masterpiece, the artist is not satisfied with the final product. This holds true for the performing arts as well. An actor or dancer is constantly seeking perfection and striving to correct the tiny flaws undetectable by the observer. The more thinking person blames himself for falling short of his goal, the more doing tends to excuse himself for turning out a second rate product and claims it is better than it is, or was bungled by someone else. It is never his fault.

There is a fundamental difference between the more thinking person and the more doing. The desire to give, to take and share affection is a powerful need of both, and is closely linked to thinking and doing. A more doing person is likely to be affectionate in the physical sense, wanting to touch and be touched. The thinker tends to avoid physical contact and thrives on being alone. It doesn't mean he lacks the ability to love and give affection, his need and capacity are just as strong, but it is difficult for him to be demonstrative. Doers find this hard to understand, just as thinkers can't understand the desire for hands on love.

The link between the thinking and the doing is not well understood, but they work together smoothly and efficiently most of the time. Thinking is controlled by the chemistry of the brain and doing by electrical impulses. It is interesting to note that doing is located in visible outgoing ways, — electrical impulses — whereas thinking is located in hidden, inner recesses—chemical changes.

There is a suitable balance for each of us. We have opposing sides, play and work, which are mostly doing and thinking. There is physical play like sports and jogging, intellectual play, such as chess and bridge. There is sexual play and addictive play in drinking and drugs, and there is emotional play, like acting and dancing. The work side has the same, and we may choose a physical occupation, as in a factory or as a carpenter or farmer. It may be as an actor or a musician, which are emotional occupations, or using the mind, as in study or research, or just to enjoy life in drink and sex.

It takes effort to reach the right balance. If you are one sided, as most of us are, it is easy to slip into using your dominant side to the neglect of the other. If you are more doing, you like to be with people and have fun and may let play interfere with your job. If you are more thinking, you are apt to indulge in solitary pursuits like reading and writing, and soon retreat more and more into yourself.

Whatever you choose to do fulfills you because it is your strongest side, and you build it up more and more, resulting in reducing the strength of your opposite side. Then there is the imbalance we mention so often, and this makes a dissatisfied person. Work for a better balance, it cannot be perfect all the time, but if you allow your life to stay lopsided, the workaholic will burn out and the playaholic will be unable to find pleasure because he has done it all.

WHAT IS INTELLIGENCE?

Intelligence is the capacity for thinking and learning, the ability to understand, to grasp concepts, to reason and to make use of and apply this knowledge. It involves problem solving skills, both mental and physical, and the aptitude at translating thoughts into words for communication with others. Intelligence is inborn and remains relatively the same throughout life.

Intelligence differs from one person to another in potential. It takes many forms, verbal, mathematical, artistic, mechanical, athletic, financial, sexual and others. We are apt to disparage types we lack. The athlete, for example, doesn't think much of the scholar, and the scholar looks down on the athlete. But they are equally intelligent in their fields.

Historically, science has concentrated on verbal and mathematical aptitudes in determining intelligence levels. Now other forms are recognized, for example, the gifted mechanic who has a dismal academic record but who excels at his trade. This takes a larger view of the individual's contribution to society, and is a sign of progress, without referring to the monetary value of a given activity.

WHAT IS ENERGY?

Every living thing is an energy transfer mechanism. What does that mean? Energy is the force that comes from the brain and is entirely under its control. It is physical and mental energy operating as one. We also call it "drive". As one of the four cornerstones of personality, energy flows from the brain into the body and directs doing. Man transforms his energy into technology which has become indispensable to society.

Each tiny bit of life, whether it be an amoeba, a plant, an animal or a human being, starts out with a burst of energy. The seed or sperm uses its little bit to look for and implant itself in the immobile egg. This is true of pollen carried by a bee from one flower to another, the seed from a tree fluttering to the ground, or the sperm of fish and animal life, to which we belong. Once it finds its niche, it takes in nutrition, turns it into energy, and grows and changes itself and its environment. It repeats this over and over until it has reached its full potential, animate or inanimate. For instance, the goldfinch converts his energy into flight and song, and joins with his mate to produce the next generation of goldfinches. The seed from a tree is nourished by the sun and the rain and draws nutrients from the earth to grow into another tree. This is true of all forms of life on earth.

The earth also is an energy transfer mechanism, acquiring most of its energy from itself and the sun and rain. It is growing and getting hotter and transforming its surface. The sun is pouring energy onto it, creating storms and winds, rains and blizzards, tornados and hurricanes.

A leaf starts as a tiny bud and takes in sunlight, water and food from the tree's roots. It grows and fulfills its mission until the fall when it shrivels and falls to the ground, a waste product to us. But to nature it is not waste, but a collection of atoms, the building blocks of matter, which are constantly changing, but always available. Our bodies are made of atoms that came from trees that may have died millions of years ago and rocks what were on the moon, or from solar dust. Our atoms will eventually be a part of some totally unrelated thing, perhaps an elephant or a tree, an airplane or a satellite.

To nature, recycling its atoms is merely a reshuffling of the infinite numbers of an ever available supply, each having a special purpose in the scheme of things. This interpretation is humbling, the orderly immensity of the plan mind boggling. This is really what physical life is.

What it adds up to is that each of us is an energy transfer mechanism composed of billions of atoms constantly taking in and throwing out energy, doing his or her thing in his or her way, always different from the next person, never two alike, a tiny part of a huge energy transforming process, each performing according to his or her personality.

Why don't people stop and do nothing for a change? Because their energy has to be used. Some move slowly and some fast and some spin their wheels, but none stands still. So the universe is in a state of perpetual motion. The earth keeps turning and living things on its surface race to use up the energy they generate, and we are caught up in this restless activity. It is the way it has always been and always will be.

Where does energy come from? From two sources, inheritance and the food, water, air and sunlight taken into the body which converts it into many activities. These may be thoughts, acts, emotions, and spiritual matters. Inborn energy levels range from high to low, and each of us differs in the ability to change it into achievement. It is influenced by changes in mood, and depleted by poor nutrition, illness, abuse of the body or injury. Energy, with intelligence, regulates vitality and persistence of effort. For instance, self-motivated work, as that of a writer or artist, takes a lot of energy. So the pace is slower if there is no outside stimulus.

If a man has a low energy level, he may be labeled as lazy. This is unfair. He will always have limited drive, mental and physical, but there are advantages, for while he may move slowly, he is more persistent than the high energy person, and less hasty in his reactions. He makes shrewder and more accurate decisions, because he takes time to absorb and consider the facts before he acts. He may live longer, because he develops a lifestyle that is moderate and easy going, and he will be happier than many others because his lifestyle is better balanced. He lives in a quieter, more self-possessed way than the high energy person, and has made a comfortable adjustment because he doesn't press himself. He is good company and makes satisfying relationships. He is a contented person, and accepts the fact that he is not a world beater and is not envious of those who are.

The low energy person with high intelligence is not a happy person, because he is always blaming himself for not getting anywhere. This creates tension and anxiety as he tries to accomplish what he lacks the energy to do, and moves from one project to another without making progress in any.

The majority of the population has moderate energy levels, and they steadily weave their way through life with relentless determination, carrying the bulk of the nation's work load. They are blue and white collar men and women, mothers and housewives, and those in the work force who contribute enormously to business and industry. There is nothing exciting about them, they are most of the people we know, and the world needs them.

Those with high energy and high intelligence are the greatest achievers, and are often in the public eye. They are presidents of big corporations, prominent politicians, famous doctors and lawyers, renowned artists, men and women in every discipline. They dash through life, always projecting ahead and planning projects and changes. The world needs them too, for they promote progress in every field.

There are high energy people with moderate intelligence. A housewife and mother expends most of her energy on family and home. She is also active in community affairs, contributing her talents to the church, the hospital or a charity. she has many projects going on that improve the home and please her family, such as knitting, sewing, cooking and gardening. Those in the work force with high energy are generally superior employees, and accomplish more than their colleagues. The exception is the person with low intelligence who is running all the time but getting nowhere. Each, however, has a place in the scheme of things.

THE PACE OF LIFE

Everybody goes through life at a different pace and at a rate best suited to the individual. Some go at a tremendous speed, some moderately, and others slowly. This is not fully understood, but pace seems to have a rhythmic pattern and is related to energy level, personality, upbringing and habits formed. For instance, the butcher sets his pace for slicing and chopping at a speed he can handle, the taxi driver doesn't have to use as much energy, so can travel faster. A low energy person growing up in New York City sets a faster pace than one who grows up in Great Falls, Montana. The president of a large American corporation sets a faster pace than the leader of a mountain tribe in Afghanistan, although they may have a similar personality makeup.

An intense, high energy person who goes at things slowly and painstakingly shows an extreme thinking nature, so he establishes a rhythm that becomes difficult to alter. In essence, he is in a rut, but can get out of it if he wants to and is willing to work at it. On the other hand, some people set a pace so fast they don't function at their best, and jump from one project to another without finishing any. This is a sign or impatience, linked to immaturity and a low tolerance for frustration. It is also related to high mood, and carries over into relationships. Such persons race through the day and don't have time for anyone, flipping from person to person,

knowing no one really well and creating unsatisfactory relationships.
It takes concentration and effort to regulate the pace of life, and the key is the determination to change. In order to slow down the pace, set priorities. Too slow a pace can be speeded up by committing to a daily schedule. The inability to change is linked to rigid right brain thinking.

ADDITIONAL CHARACTERISTICS
There are other personality traits which stay in the background waiting to be called upon when needed. They are passiveness and aggressiveness, which are opposite forces, and obsessive compulsiveness. All are closely connected to thinking and doing, and are not noticeable until a challenge awakens them. A passive person is submissive and unresisting, and doesn't react to events that stir up emotions. An aggressive person starts unprovoked attacks. There are varying degrees between these extremes, and how they are used depends on the circumstance that aroused them.

Most people have degrees of these traits which may be low in one and high in the other, with a bit of the opposite showing occasionally, or they may range widely. It is hard to predict how far a person will go. Someone not normally aggressive may surprise himself by an overreaction when really angry, while an aggressive person may face the same situation with uncharacteristic passivity.

Compulsive obsessiveness is our fussy quality and shows up at an early age. Some babies can't stand to have wet faces, cry until their mothers wipe them and scream until their diapers are changed. Other babies are not bothered by wet faces or dirty diapers and go happily to sleep, totally oblivious.

If this trait is strong, we keep tidying up around the house and at work, and when something is out of place, we are uncomfortable until we put it where it belongs. We are meticulous about appearances, and disorder is distressing to us. Those who have little are untidy about their persons and belongings and unconcerned if the house or office is in shambles. They don't care what others think and can be comfortable in disorder. Most of us fall in between these extremes.

Some people are so compulsive that they pick up every thread and tidy the house over and over. Some brush their teeth or wash their hands repeatedly, or carry out some pointless activity again and again. This is the extreme form of obsessive compulsiveness and is neurotic when it interferes with one's life. More about this later.

3 NEUROTIC TENDENCIES AND NEUROSES

What is a neurotic tendency? A neurotic tendency is not a neurosis or a mental illness. It is a natural function of the brain and we all have them. Being neurotic is the same as having a neurotic tendency and most people are neurotic at some time. While it is a distressing feeling, it is not serious and will pass with or without help. Those most prone have strong thinking or doing characteristics colored by paranoia. Heredity may also be a factor.

A neurotic tendency is a premonition that may develop into a neurosis, and we can have them repeatedly without causing symptoms or becoming a neurosis. For example, Veronica, a professional woman, lives comfortably with several. She has a "thing" about snakes and squirmy creatures, she is afraid of heights and at times suffers from claustrophobia, the fear of being closed in. Each came from a childhood experience. She was bitten by a snake, she fell out of a tree, she locked herself in the bathroom and couldn't get out. These incidents set patterns in her impressionable child's mind which emerge when she is under extreme stress, but none has turned into a neurosis. She handles these tendencies by avoiding situations that provoke them.

What is a neurosis? A neurosis is when a neurotic tendency becomes so severe that it interferes with your life. It is not clear why it occurs, however, it can be treated and cured. Usually an imbalance takes place between thinking and doing, and if you have strong qualities, you are more susceptible and the struggle is intensified.

For instance, Elizabeth has an aversion to open spaces, a neurotic tendency, which she handles comfortably. She is the mother of eight which requires spending most of her time at home. Thus, over the years her doing became subordinate to thinking and, when the children grew up and left home, her phobia increased to the point that she couldn't leave the house alone. Her thinking and doing had grown out of balance, and her ability to make an adjustment was blocked by the struggle between them. She has a neurosis.

Going into a neurosis is like going into a downward spiral. You do less and less as you go round and round descending and relinquishing your obligations one by one until you reach the bottom. There, you are unable to move. You are in the deep pool in the middle of the lake. You hate yourself for shirking your responsibilities. You think rationally but can do little, and this bothers you the most. You go to the office but can't seem to work, you eat poorly and are ashamed of your appearance. What to do? You know that you have to climb back up the spiral and resume your life, but you need help. Where do you go? If you don't know a psychiatrist, ask your doctor, if you don't have a doctor, call the medical society or the hospital. Through therapy, you will soon be able to start living again, and little by little, the balance between thinking and doing will be restored.

What tips the scale into neurosis? When a person's health is at its best, every system in the body is in balance. We call this "homeostasis". When a system gets out of order from an illness or a shock, an imbalance follows, and the healing forces in the body go to work. Most of the time it is successful, but there are some people who react to stress in such a way that they slip into a neurosis. Why? Such people have inherited a tendency to over or underreact to stress with depression and anxiety. As they face stress over and over again, they learn that depression and anxiety follow, so they react before stress occurs and develop a neurosis.

Neuroses are elusive and turn up in different forms.A neurosis shackles you and you feel paralyzed. You worry needlessly. Your son is going to flunk out of college, your husband is overweight and must have diabetes, your father has heart trouble and is going to die. Is nothing right? Why are your thoughts so sick? Because your thinking and doing are unbalanced. Can you take something for it? No, drugs don't help neuroses, except temporarily, but patients often dose themselves with over the counter remedies, and worse, whatever is in the medicine cabinet. They take tranquilizers for anxiety, alcohol for depression and consult priests and rabbis for phobias. All to no avail.

Don't blame yourself if you succumb. Life is not easy. It offers four major challenges that make constant demands; getting along with others, succeeding in your work, adjusting to a fast-paced society, and being satisfied with yourself.

These create self doubt about money, sex, and personal relationships, and may be compounded by the loss of a loved one or a job, divorce, or an illness. These problems are fertile ground for anxiety and depression, and you may feel at the limit of your tolerance.

To make the correct diagnosis of a neurosis is an art as well as a science, and many experienced psychologists and social workers are exceedingly capable at this. Someone, however, has to take the responsibility for the diagnosis, treatment and outcome, and this should lie on the shoulders of the psychiatrist. The non-MD. runs the risk of missing the brain tumor, the thyroid disturbance, the endocrine abnormality, the psychosis or the potential suicide. All can be mistaken for a neurosis.

The boomerang affect

Recent studies prove that the Western world is more and more aware of anxiety as a common condition. The technology of the modern environment is a psychic boomerang. It has made living more convenient, comfortable, and interesting, and there is more leisure time. These are the pluses, what are the minuses? The rapid pace of life, the complex demands of science, and social dissent are only some of the pressures that double back on us.

We are confronted with the sexual revolution, drug abuse, the threat of AIDS and increasing crime. Civil unrest is interspersed by waves of terrorism and the moral fabric of our society is eroded by revelations of fraud and deception at every level of business and government.

The influx of immigrants, their different cultures intermingling, only seems to complicate things. This was supposed to enrich our country, as it did in earlier days, but each culture seems to weaken and become unrecognizable, choosing the lowest common denominator in American society. As a result, no one is comfortable in his own milieu.

The world has changed, society has changed, but man has not. He has the same mind and the same body, which has not been modernized or mechanized or conditioned to be computerized. Is it any wonder that neurotic tendencies develop into neuroses?

THE NEUROSES
Anxiety

Normal anxiety is the feeling of tension, apprehension, indecisiveness, foreboding rolled into one, and is a natural reaction to challenge. Anxiety is constant and flares up as we meet minor and major altercations: a flat tire, the angry retort of a friend, spilled milk, the first day on a job. Watching a child board the school bus can create anxiety, so can catching a train or a plane. In fact, anxiety is an essential ingredient for meeting life's challenges and obstacles and a major component of drive.

Anxiety is universal. We can learn to deal with it and put it to work for us, and properly used, it enhances life. It loses its usefulness when disregarded or masked, as by tranquilizers. Anxiety is like apprehension and is linked to the healthy fear of paranoia, which keeps us out of danger. Similarly, it helps us do a better job. When we are anxious before an interview or a performance, studies show that apprehension results in a superior production. Anxiety keys up mind and body for a special effort.

Society sets goals for everyone. Thus, competition intrudes on every phase of life, and competition is a source of anxiety. We compete in social, personal, recreational, financial, and sexual areas, to name a few. Top performance is so entrenched in our culture that self-esteem is directly related to success, and constant striving toward goals beyond our capability further produces anxiety. Is anxiety the same as stress? No, but they are related in that each causes the other. Stress comes from without and anxiety from within. Stress, in the following example is the anticipation of performing, causing the anxiety within. Three of us are in a dance competition waiting to go on. One says: "Why do I get myself into this?" The others concur, but here we are, tense, stomachs in knots, hearts beating fast. This is normal anxiety.

However, anxiety has another face, neurotic anxiety. It is when normal anxiety becomes chronic. It can be severe following a devastating life event, but if it continues without let up and interferes with normal activity, it is neurotic, then professional help is advisable. How do we recognize neurotic anxiety? Symptoms vary, but the usual pattern is a feeling of helplessness and inability to reason, and an unfounded sense of foreboding and danger, in short, a feeling of "going to pieces". Negative thoughts and fears of not being able to cope invade the mind. The slightest crisis seems overwhelming and the future holds no hope.

An anxiety attack is a mild form of panic attack. The person is afraid to leave the house, afraid to drive the car or do anything that might trigger off an episode. Neurotic anxiety digs up bad memories and builds negativism. "I was unsuccessful at that, so I will fail again". Physical symptoms appear, such as muscle weakness, stiff joints, trembling, increased heart rate. Trying to control these warnings by denial does no good, or plunging into hard work, oversleeping or eating, drinking or drugs. Anxiety increases.

What makes this happen? Recent studies show that the tendency toward neurotic anxiety is an inborn quality shared to some degree by everyone, and the more thinking person is the most vulnerable because of his sensitivity. Fifty years ago it was believed that it stems from childhood experiences. Parents and relatives may tease a child for buck teeth or protruding ears, or belittle his performance. The peer group also plays a part, and insensitive criticism can have a detrimental effect. However, all childhoods have trauma but not every trauma results in neurosis.

It takes a skilled psychiatrist to diagnose neurotic anxiety correctly, and this is essential, for there are physical illnesses that show the same symptoms. It is also necessary to have a thorough physical examination, including laboratory, x-ray and other tests, before a diagnosis is attempted. Anxiety attacks are easily treated with one of the benzodiazepines, which are not heavily sedative and help a person relax. They seem to supplement or replace the endorphins in the brain, which are our natural anti-anxiety chemicals, and are not addictive. Psychotherapy as an adjunct to the drug, helps the patient understand his illness, restores self-confidence and self-image, and rebuilds emotional equilibrium. The length of treatment depends on the severity, depth and duration of the neurosis which can be shortened considerably by vigorous daily exercise.

Insecurity

The development of security in an individual depends on the fulfillment in childhood of three needs: love, approval and consistency. Parents are especially important, their love is as vital as food and shelter to the development of a child's personality. Parental acceptance of children as they are, and consistency of love and approval helps them emerge as secure adults. Parents who offer this support ensure emotional security.

Years ago, babies in institutions were left alone except for regular feedings. They were not picked up when they cried, not fondled or played with. In the absence of affection and physical contact, many lost their appetites and died for no reason that could be determined. Subsequently, studies show that babies who get tender loving care flourish and grow into happy and healthy children. This led to the development of surrogate mother programs for orphaned and abandoned little ones.

We all feel insecure at times. It is an unpleasant uncomfortable emotion, and even if our family is sympathetic we still feel undeserving and uncertain. The best remedy is the love, approval and understanding of those dear to us. Attention and admiration from other sources can also boost the ego and help us meet the demands of the job and social obligations despite a sense of inadequacy. But we must believe this approval to be genuine, not superficial flattery.

Insecurity, faced with severe emotional stress, can get out of hand and become a serious neurosis, provoking us to turn to some chemical like alcohol, drugs or medication. We might join a religious cult or bury ourselves in work or recreation to the neglect of our family and friends. Whatever we do we do in extreme.

So, we are all susceptible to a neurosis in the face of prolonged and painful stress. When it occurs, neurotic tendencies are accentuated. If we fear heights, we fear them more, if we are subject to air hunger, we have more attacks. If all this becomes unbearable or interferes with our lives, professional help is indicated with an excellent outcome in almost every case.

Depression
Depression is our state of mind when the mood is low. We all are subject to swings of mood of varying heights and depths. When mood is low, it is depression, when high, it is elation. Some of us travel on an even keel, some have minor ups and downs, while others experience deep valleys of depression and heights of elation. Some have high moods and are never depressed, but a more common pattern is depressed episodes, then mild elation. All mood changes are related to internal and external pressures. Those from within refer to the state of our mental and physical wellbeing, and those from without to our circumstances at that particular time.

Medical science is discovering more about depression. The seasons of the year have an effect, and mood can plummet under adversity, tragic life situations, accidents or surgery. Often we blame the weather, but contrary to what most think and many believe, studies reveal no relationship to gloomy days, barometric pressure, the phases of the moon, tide fluctuations, the positions of the stars or other natural phenomena.

Are there different kinds of depression? Yes, two types. "Reactive" depression, the most common, is described here, and the psychotic types later on. Reactive depression can result from something that happens to a person emotionally or physically. If emotional balance is disturbed, such as by the death of a loved one, a Job loss, a broken love affair, a financial setback, depression may set in. Repressed anger can also find its outlet this way. If a person's physical balance is upset such as by an accident, a serious illness or incurable disease, depression can be the response. How we react to such pressures is individual, and what depresses you may not depress me, or to the same degree and vice versa.

Since mood swings up and down, certain events can send it very high, Extreme elation is just as disturbing to emotional balance as deep depression, and actually is more harmful, for it can cause compulsive behavior that produces guilt and then depression. For example, Arthur wins a million dollars in the sweepstakes. He is euphoric! He buys a Rolls Royce, sells his house, buys another and spends a fortune on furnishings. He entertains lavishly and gives expensive presents to his family and friends. Soon the money is gone and he sinks into the depths, infuriated by his foolishness. Although it is hard to believe, people do react this way.

Abnormally high or low moods are caused by fluctuations of the chemical balance in brain cells. We all experience them, and knowing they are shared alleviates our sense of hopelessness. Fortunately, swings in mood are temporary, and they vanish without a trace.

Reaction to a life crisis can result in complaints other then depression. Headache, fatigue, and vague pains are common, also nausea, constipation, heartburn, and irritability. A person may gain weight due to compulsive eating, drink more, smoke marijuana, take tranquilizers, or resort to hard drugs, in an effort to escape. This is the body's attempt to ward off depression by substituting physical sensations, and may succeed temporarily as a sort of anaesthesia. However, there is still a risk of depression, especially if alcohol is used, because alcohol is a mental as well as a physical depressant.

How do we feel when we are depressed? It is difficult to concentrate and productivity drops below normal and may become noticeable on the job. Energy and interest decrease, we lose initiative and self-confidence and low output alarms us. We may lose interest in sex. When we awaken, we dread facing the day and don't want to meet people, even family. We take a long time to dress, and look in the mirror with distaste and wish we could crawl into a hole. It takes a major effort to leave the house and seeing people is exceedingly painful. This suffering is as valid as that of physical pain, such as a broken leg, or major surgery. Though we may not believe that it will pass, the pain is easier to bear if we can accept that it will. The symptoms of reactive depression are similar to those of chemical depression, one of the psychoses. More about that later. Only the most skilled physician can tell them apart. The correct diagnosis is important because the treatments differ, reactive depression is treated by psychotherapy, chemical depression by drugs. Furthermore, reactive depression comes and goes, whereas chemical depression is static and much deeper.

What can we do about reactive depression? Understand yourself, know what affects your mood, and maintain physical well-being. Try to bring the extremes of high and low moods closer together by easing the pressures around you. Do this by showing your personality, and your emotional, intellectual and physical capacities associated with work, family and social life. The thinker suffers more from anxiety than the doer, and a high energy level can drive you beyond your capability, so it is important to set goals.

If you haven't recovered after a reasonable length of time, you need better understanding through psychotherapy. As there is always a cause for reactive depression, it is important to tell the doctor the recent events of your life as well as your physical and mental symptoms. A good rapport, with open communication, is essential in making a diagnosis and to benefit from therapy. The next step is to listen to your doctor, although his suggestions may be difficult to accept. He may recommend that you change your surroundings, perhaps get a new job, take up a new sport, move to another town. Even though you feel as though you will never be your real self again, believe your doctor when he assures you that you will. Then you and he can work out together the changes in surroundings that will relieve your depression.

During this time it is important to maintain your health through good nutrition and exercise. Adequate sleep is essential and curative, and you may require tranquilizers in the beginning, but a more effective treatment is psychotherapy without medication. Your well-being is permanent when it results without chemical support.

HYPOCHONDRIA, THE MEDICAL TERM IS HYPOCHONDRIAS

The hypochondriac is a person who is always complaining of physical symptoms. He is absorbed in them and talks endlessly about them. While they are usually non existent or greatly exaggerated, the hypochondriac does not feel well, and his pains are very real. Despite his complaints, however, he functions well in his work and in other aspects of life.

Hypochondria is a neurosis common to both men and women, and is classified into two groups. Those who make up the largest are afraid a serious condition is causing their symptoms, and that their doctors are going to find a fatal disease. Their paranoia is exaggerated. A smaller group are persons with a strong paranoid tendency who use complaints of illness to evade responsibility. They learn early in life how to manipulate others by their moanings and groanings, and to avoid doing things they dislike. They use their pains as an excuse not to work hard, with the added bonus that people feel sorry for them and give them the attention they crave. This kind of maneuvering becomes a way of life, and some get away with it and manage to make a living, taking all the help and attention they can. Others lose the respect of their friends and wear out the patience of their families.

The hypochondriac usually develops physical symptoms by early adulthood. Most have vague, transient aches and pains during childhood, described as "growing pains". They may be chronically tired, lethargic, irritable, and talkative. Sleep is disturbed by nightmares. While these signs are intermittent during childhood, a pattern becomes established by the time they are grown up.

Hypochondria is poorly understood, the underlying cause unknown. It is clear, however, that life situations combined with certain personality traits are partly responsible, and it may be theorized that a traumatic experience, a depression, or a feeling of inadequacy could set it off.

Another explanation could be the underproduction of endorphins in the brain, the chemical that provides anaesthesia for pain, making some people overreact to the minor pains and distresses we all have. We just don't know. There is usually nothing physically wrong, but the hypochondriac's paranoid tendency never allows him to accept this, so he seeks relief from one doctor after another. In essence, his complaints come from his mind, making it difficult to diagnose and treat him.

Assess your body at this moment. You probably have mild discomfort somewhere, an itch, your toe hurts where you stubbed it, your back aches from sitting, your knee cracks when you bend it. But you don't dwell on these annoyances, you brush them aside, never thinking you may end up with some disease. The hypochondriac's pain, however, is not mild. To him it is major and bothers him constantly. And he is honestly afraid.

Men pay little attention to signs of something wrong, but it is the nature of women to be aware of what is happening to their bodies. Women seek medical advice more promptly than men, who prefer to enjoy life and want to keep it as it is. They are unwilling to upset the apple cart by zeroing in on something they consider unimportant. There is a middle ground, however, between hypochondria and burying your head in the sand.

What are the symptoms? They can be many and confusing and there is no set pattern. Each patient's complaints differ, and no two doctors agree on what they have been told.

It may be weakness and fatigue, or vague aches and pains moving from place to place. There are gastrointestinal symptoms, irritability, insomnia, and chronic physical exhaustion.

After a day's work, a change of pace is reviving to us and we can enjoy the evening. Not so the hypochondriac. He feels tired when he gets up in the morning and tired all day. An evening of fun doesn't entice him. He is too tired and too sick to go anywhere or do anything for pleasure, or to indulge in sports or exercise. He bemoans the fact that he can't, yet seems to revel in his helplessness. This is a human trait and we all have a trace of it at times. Some hypochondriacs have low self-esteem, and their symptoms give them a sense of identity which boosts their egos, so they hold on to them and use them as a crutch to prove they are somebody. Attention relieves their guilt, but this only perpetuates the condition, which they deny is psychosomatic, for that to them is demeaning. A physical disease is acceptable and an excuse to duck responsibilities.

A hypochondriac visits one doctor after another, looking for the answer he wants to hear, that be has a disease. If he admits how many doctors he has seen, the picture would be clear, but no physician will make a diagnosis on that evidence. So the round of tests begins anew to make sure there is no organic cause. This focuses attention on the patient, which is just what he wants. Sometimes he says he is feeling better, only to report the next day that he is worse. This is frustrating to the doctor who is often blamed for adding to his suffering. Moreover, he takes pleasure in making it hard for the doctor by cancelling tests, refusing to take medication and being generally uncooperative.

If the diagnosis is hypochondria, and the doctor explains that the symptoms are imaginary, the patient becomes angry and sets out to find another doctor. This again is paranoia. The hypochondriac is delighted if the doctor turns up a reason for his symptoms. It may be an ulcer or gallstones or high blood pressure, but now he has something to blame and would cheerfully undergo surgery. In fact, he might demand it.

What kind of person becomes hypochondriac? These people are victims of their personalities. They have low energy, average intelligence and inclined to be the doing type. They come from all walks of life and all types of occupations. They have difficulty dealing with life's problems and are unable to accept others whose personalities differ from theirs. Most function fairly well. An example is "Old Bones, an outstanding ball player. He comes to the ball park announcing that he feels terrible, then proves to be the best player. He complains but always outperforms the others. This is typical of the hypochondirac without a paranoid tendency, with a paranoid tendency, he would use his symptoms to get pity and avoid playing.

It is difficult to diagnose hypochondria. The symptoms are vague and if the patient has been making the rounds of physicians' offices, it is easy to assume that he is a hypochondriac. However, a physical condition can account for the complaints, therefore a thorough workup followed by a psychiatric examination is indicated. Prior to this, a diagnosis of hypochondria should not be considered.

Here are examples. Thomas complained of severe pain in his neck. After a workup, exploratory surgery on his spinal cord revealed nothing. The pain persisted, and the surgeon, suspecting hypochondria, sent the patient to a psychiatrist, who treated him for depression, but could find nothing else wrong. However, the pain continued, so he sent him back to the surgeon. A myelogram revealed a block in the spinal canal, and subsequent surgery disclosed severe disc disease.

Edward went to the emergency room with severe pains in his chest and shoulders and muscle tension in his neck and back. An electrocardiogram was normal, and he was referred to the psychiatric clinic. Instead, he went to his medical doctor, who hospitalized him and called in specialists. They could find nothing wrong, and he was referred to a psychiatrist who diagnosed endogenous depression, prescribed a drug and the symptoms soon disappeared.

So there is no specific method for treating hypochondria. Each patient is evaluated on the basis of personality, history, and life situation. Psychotherapy is successful in many cases, and drugs in selected cases. Tranquilizers have little effect, and shock treatments make the symptoms worse. The doctor may recommend changes in lifestyle, job, personal relationships and so forth. If the patient is agreeable, there is a good chance his symptoms will improve. If he is not, his chances are slim.

Controlled studies show that there are trends in hypochondria. In other words, fashions influence the types of complaints. Lower class groups show a higher incidence of psychosomatic illness than upper class, and are less responsive to treatment. These patients reject the possibility that personal problems cause their illness, and expect the doctor to cure them without cooperation on their parts. Upper class patients can usually be helped once they have accepted the fact that a physical condition is not responsible.

HYSTERIA AND HYSTERICS

There are common misconceptions about hysteria and hysterics.

Hysteria summons up a picture of a woman weeping uncontrollably, screaming and throwing herself on the floor. This is not hysteria, she is having hysterics, which is an over-emotional reaction to stress or frustration. Temper tantrums in children are hysterics, and the child who holds his breath until blue in the face. Hysterics is a form of manipulation (an element of paranoia) and a frantic call for attention or to force someone into doing his will. Men, who are inclined to keep their emotions within and become withdrawn when in distress, are less prone than women to hysterics and become sullen and speechless instead. Some women react by "going into hysterics" and throwing their bodies about, pretending to hurt themselves. They like to give that impression but are seldom hurt. So hysterics is not to be confused with hysteria.

Hysteria is a neurosis. The medical term is "conversion" hysteria and it is a primitive response. Hysteria is a product of the mind, causing the body to convert emotion into physical signs. It affects women who are usually of the more doing type and is rare in men. Those who are prone react primitively in other areas as well.

Its symptoms mimic hidden desires and are dramatic, like excitability, histrionics, self-centeredness. They are colored by every element of the personality and thus differ from person to person. There are women who experience false pregnancies, with enlargement of the abdomen and cessation of menstrual periods. Tests prove these cases to be hysteria triggered by an intense desire to have children and the failure to conceive.

Sudden paralysis is not uncommon in hysteria. Invariably, the symptoms show up in the part of the body related to the cause. For example, Helen suddenly developed paralysis of the legs and pelvis. She had not been injured and physical examination revealed no cause, so a diagnosis of hysteria was made. Psychotherapy revealed a broken love affair. The paralysis was in the site of her frustration, the sex organs.

Patricia slapped her boy friend in the face. He left in a huff and immediately, her hand became paralyzed from the wrist down. Frightened, she want to the emergency room and was hospitalized for tests. When no physical cause was found, she was referred to a psychiatrist who treated her by psychotherapy. The paralysis disappeared in two days.

Hysteria is the great imitator. It takes many forms and resembles symptoms from many diseases. A woman may suddenly become blind or deaf or go into convulsions. Surely, this is an epileptic attack. But no, it isn't. So the physician can make a diagnosis only after thorough examination and testing.

Another form is amnesia. A woman goes out of the house to shop at the supermarket. Suddenly, she doesn't know where she is. Then, she finds herself in a strange location with no knowledge of how she got there and doesn't remember her name or where she lives.

The woman who has hysteria is emotionally unstable, and in many cases gives the impression of seductiveness, which may or may not be intentional. She is inclined to be self-centered, and sometimes overreacts to get attention. She can be attractive, however, in a social group but is usually insecure, which leads to anxiety so severe that hysteria provides her need for attention.

An hysteria patient has no concern about what seems to be a serious illness. She may become blind, deaf, or partially paralyzed without showing fear. In a patient who claims to be severely ill, this is a leading clue for the physician. A common mistake is to suspect a physical disease.

Hysteria used to be fashionable. Historical novels show instances of hysterical paralysis in women, which disappears when their wishes are granted. In olden days, it was considered romantic for a lady to faint in her lover's arms. Someone ran for the smelling salts, and a whiff quickly aroused the fainting beauty. We don't hear of women fainting today.

Hysteria is contagious. We see a crowd and hurry to find out what it is about. Is there a fight? Is someone hurt? We have to know. At the scene of an accident, everyone stands around and watches, but only a few help. We have all been held up on the highway by rubber necking. Go into a store, stand at an empty counter and stare at something in the showcase. Pretty soon three or four people join you!. This shows the "catching" element of hysteria. We follow one another like lemmings.

There also is mass hysteria which sometimes occurs when people gather together under conditions that are stressful. One person becomes ill, then others, but not all. Symptoms may be fainting, headache, difficulty in breathing, nausea and vomiting and doctors are called in. If physical examinations and laboratory tests eliminate medical causes, the diagnosis is mass hysteria. It is a syndrome in which the emotions stimulate the body to produce symptoms, sparked by a neurotic tendency and inflamed by paranoia.

These phenomena can be caused by public concern about toxic wastes and can provide fertile ground for an outbreak of mass hysteria. This has occurred in towns that were sprayed with insecticides. Dizziness, headache and general malaise showed up with symptoms ranging from mild to severe and incapacitating. None of the victims had organic disease and complaints disappeared as fast as they came.

Recently, a manufacturing plant was shut down because of a complaint of bad air. An employee felt sick and blamed the air, another succumbed and another until most of the workers claimed illness and the plant had to close for the day. The air was not bad. This is mass hysteria, and the emotion that makes it spread is fear. Is this going to happen to me?

Mass hysteria is suspect if the victims are normally healthy, if there are neurotic signs such as fainting and hyperventilation, if the illness spreads quickly and clears up as rapidly as it appeared, and if it reappears when the same individuals return to the first site. There are no studies that show which personality types are vulnerable to mass hysteria. Usually, up to forty percent is afflicted, more women than men in a ratio of about sixty to forty. Young people, especially girls, are vulnerable, and sixty percent of recorded events occurred in schools, often coinciding with stress associated with examinations, sports competition or performances.

How do you treat hysteria? Psychotherapy and hypnosis are the treatments of choice, separately or together. Drugs and tranquilizers do no good. Prognosis is excellent.

OBSESSIVE BEHAVIOR

Do you find yourself counting as you walk up stairs? Do you avoid the cracks in the sidewalk, count the cars passing by as you wait to cross the street? Do you notice how a man keeps arranging the articles on his desk as he talks to you? Or a woman is always tidying up and wiping off the kitchen counters over and over? Do you find that you develop unconsciously little tricks which become a part of you? Do you become so involved in an activity, such as golf, tennis, fencing, that you can't put it out of your mind? These are manifestations of obsessive behavior that everyone has. They are not neurotic. They are the idiosyncrasies that express your uniqueness. If you are the thinking type, you show more obsessive behavior due to your concern about details and the observance of formalities. You spend minutes organizing the articles on your desk and arranging the papers in neat piles and concentrate on these to the neglect of important things. When you are finally ready to work, you focus on the minor areas and delay the major. All this is normal, especially for the more thinking type of personality, but can get out of hand if not watched and controlled.

Occasionally, obsessiveness builds up until it begins to interfere with the daily routine. We spend so much time satisfying these needs that we neglect obligations, even though common sense tells us to stop. Our behavior is no longer normal. It has progressed into a neurosis. Now, we need help, and psychotherapy offers a favorable outcome.

There are three types of neurotic obsessive behavior. One, indecision, introspection, brooding, bewilderment. Two, obsession to create and carry out rituals. Three, terrifying thoughts such as the temptation to do away with someone.

One: You can't make decisions. Your mind flits from the pros to the cons, doubts crowd in and upset you. You are forced to make a choice, for the question, which is usually minor, seems momentous. You can't make up your mind and this upsets you more.

Here is an example. You park the car and turn off the lights. In the office, you begin to worry. Did I turn the lights off? To make sure, you go out to see. An hour later, you wonder again, and make a second trip to check. You misplace your glasses. You go to where you think you left them over and over, even though you know they are not there. Indecisiveness bothers you all day. Such flareups disappear as quickly as they came.

Two: The creation of rituals. You are compelled to wash your hands repeatedly, and exhaust yourself washing the same dishes and pots over and over. When out for a walk, you are forced to count every step, and if you miss one have to go back where you started. You make up mannerisms you have to use, like clapping your heels together and blinking before extending your hand when you meet someone, or counting the money in your wallet before paying for a purchase.

Three: Terrifying thoughts. This is the most disturbing. Suddenly, you think terrible things. You want to hurt a loved one. You even think of murder. Guilt drives you to distraction.

If you are more of a thinker than a doer, you have a tendency to become neurotic. You are a perfectionist who frets about details, your habits are fastidious and finicky. You are likely to spend hours on grooming and maintaining order in your home and place of work. You are oblivious to the fact that you are frittering time away.

You are highly intelligent, reasoning, rational and obsessed with facts. Your emotionality is low, and you don't have much sensitivity or artistic ability. You are honest and proud and have great self-confidence. You are apt to address others in a peremptory manner, for you are short on graciousness and charm. You have no sympathy for those who appreciate beauty and art, or enjoy romantic emotions. You believe in reason and cold, hard facts, and seldom allow yourself to speculate about ethereal subjects or fantasize on secret desires. Your real love is for factual enterprises like mathematics, computers and the exact sciences.

Can such a neurosis be treated? Certainly, and with favorable results. Like all neuroses, obsessive behavior is treated by psychotherapy. Drugs are of no value and should not be encouraged.

What about prevention? Science does not know how to forestall the development of this disorder. Therapy is advisable early if it seems to be worsening, for the length of time required for success depends on its severity at the time of first treatment. The cause is the same as that of all neuroses, an inborn tendency. But why one person develops it and another with a similar personality does not, is not clear.

PHOBIAS

A phobia is a neurotic tendency. Just about everybody has them. Phobia comes from the Greek, phobos, meaning fear, but phobias are not to be confused with fears like fire, flood, or accidents, for these are dangers linked to the fear facet of paranoia, which protects us. There are two groups of phobias, about things such as germs, animals, flying objects, insects, lightning; and about surroundings, such as heights, open spaces, the dark, speaking in public, stage fright, flying, elevators.

It is the nature of man to have phobias and there is no scientific explanation why. Some people are more phobic than others. Those with a strong thinking element are more susceptible, and the more sensitive they are the more intense are their phobias.

Phobias are unreasonable fears and are brought on by traumatic experiences. We can develop them at any age and the degree to which they affect us depends on our emotional balance at the time. When this is upset, phobias are aggravated. There are times when they don't bother us, and they come and go for no apparent reason, but we can link their intensity to the amount of stress we are under. Contributing factors to their development can be over indulgence in alcohol, the abuse of drugs, or excessive use of tranquilizers. They can also develop during depression.

A phobia can emerge out of the blue with no relationship to an emotional shock. However, you can always connect it with an unpleasant experience, even though it may be foreign to the phobia. For example, you speculate in the stock market and your losses are more than you can absorb. You dwell upon your foolhardiness so intensely that it creates a fear of falling downstairs, a totally unrelated concern.

Another example is Natalie, a child recently orphaned, who went to live with her grandmother, an austere, cold and strict disciplinarian. The nine-year-old was forced to run errands to unlighted parts of the house, and would race down the corridors in terror of the "spirits" that were chasing her. Her emotional imbalance from the loss of her parents created a fear of the dark which developed into a phobia.

A phobia can be catching, as Jim and his brother Ted aptly demonstrated. They developed phobias about spiders at a young age. Since small boys are rarely afraid of insects, where did this come from? It turned out that the parents employed a maid who became hysterical at the sight of a spider. Her phobia was picked up by the boys.

What are the symptoms of a phobia? The immediate reaction is fear. You become breathless and shaky and your heart feats fast. This triggers a sensation of not being able to get enough air, so you breathe harder, deeper and faster. Your heart beats more rapidly, your arms and legs weaken, you feel dizzy and lightheaded, your ears ring, your fingers tingle and you perspire profusely. Then, you panic.

You are hyperventilating. What is that? Fear touched off a mechanism whereby you take in large amounts of oxygen and blow out carbon dioxide. This causes an imbalance in the blood which brings on the symptoms. Oxygen is essential to life, and so is carbon dioxide, which is considered a poison. However, too much oxygen can be poisonous too. You need both these gases, but in proper balance.

So what do you do? One of two things: run from the phobia; exercise uses up the excess oxygen. Or stop taking in so much this way: keep your mouth shut, close one nostril with your forefinger, and breathe gently through the other. In ten seconds your symptoms will disappear. It is amazing how quickly the body returns to normal if allowed to do its thing. This is an example of how closely body and mind are linked.

Hyperventilation can be produced and as easily reversed. Take in seven to ten deep breaths. Almost immediately you will feel the signs of air hunger. Some people have a delicate oxygen/carbon dioxide balance and respond more rapidly than others. The physical distress produces the emotional reactions which, in turn, trigger the physical response. If the panicky feeling makes you breathe even more deeply, your adrenalin output increases, setting off your "fight or flight" mechanism. Panic is a primitive response that makes clear and rational thinking impossible.

How are phobias treated? There are three forms of treatment: psychotherapy, the most successful, can be tied to hypnosis or situational therapy. Hypnosis is discussed later. In situational therapy, the therapist shares the situation that produces the phobia with the patient and this gradually conditions him not to fear it.

For example, Karen was bitten by a dog and developed such a fear of dogs that she froze very time she saw one. Her distress was compounded by embarrassment, so she went for help. The therapist guided the conversation toward dogs, and soon she was able to talk about them. Then, he brought a puppy into the office and fondled it on his lap. He gradually persuaded her to touch it and she began to enjoy its cuddly playfulness. She is on the way to recovery.

William developed a dread of being enclosed in small spaces, especially elevators, into which his work often required that he go. His fear became so severe he had to plan his schedule around his phobia. Finally, he sought advice. Psychotherapy slowly made it possible for him to go with his therapist to a building and take the elevator. Gradually, he made more frequent trips with his doctor, and later with his wife or a friend. Then he could go alone. This is situational therapy.

Group therapy is popular for the treatment of phobias, and the results are excellent. Phobia centers are established around the country where fellow sufferers share problems about phobias and this helps to hasten recovery.

Occasionally phobias get out of hand, depending on the personality of the individual. The sensitive thinking person is likely to develop a phobia so severe it requires professional help, however the prognosis is excellent in all cases. So there is no reason to restrict a lifestyle because of a phobia.

We can generate a phobia about almost anything, and put a Greek name to it. Claustrophobia is the fear of closed in places, acrophobia of heights, agoraphobia of open spaces. Agoraphobia has taken on the meaning of omniphobia, or fear of everything or anything. There are people who fall into this category, whose phobias have multiplied to such an extent that they cover every situation and object, and become prisoners in their homes.

PANIC ATTACKS

A panic attack is a disabling episode of fear. It is not a neurosis, but becomes one when it occurs repeatedly and interferes with a person's life. The dictionary states that panic is "a sudden, overwhelming fear, with or without cause, that produces hysterical or irrational behavior". Symptoms appear without provocation and terror sets in. It is a frightening feeling, much like that of experiencing a phobia. Your heart beats fast, you can't get your breath, your mouth is dry, you feel weak and start to shake. You are dizzy and faint and may have pains in the chest, which suggest a heart attack. You are embarrassed because you are sure people are watching and you wonder what they think of you. You are anxious, apprehensive and afraid of going crazy, that something awful. is going to happen. Are you going to die?

There are many manifestations of a panic attack. No patient has them all at one time but may have several, and if the condition is not treated, others may develop. Hyperventilation may be the first symptom, followed by panic and fear. For example, a woman was on a train when she suddenly experienced an overwhelming feeling that she could not get enough air, so she gasped faster and faster and became dizzy and weak. She got off the train at the next stop, shaking and scarcely able to walk. She struggled to a taxi and begged to be taken to a doctor's office. She was given a sedative and quickly recovered. It was a harrowing experience. This is hyperventilation.

Our bodies require a balance of oxygen and carbon dioxide taken into the lungs with each breath. This is regulated by the autonomic nervous system, and when we breathe faster than the system prescribes, we take in more oxygen than can be metabolized to carbon dioxide, which causes an imbalance and makes the body unable to function properly. To correct this is easy. Simply shut your mouth, and close one nostril with a finger. Then, breathe slowly through the open nostril. Within a few seconds, the symptoms will disappear. Another method is to breathe into a paper bag held close over the mouth and nose for a few breaths. This prevents the intake of oxygen. Or, take a fast run around the block and you will quickly dissipate the excess oxygen.

Two-thirds of sufferers of panic attacks develop a phobia about the time or the place that an attack can occur. These are two types of locations: closed in places (claustrophobia), and open spaces (agoraphobia). Panic can take place in elevators, in church, in crowds; airplanes, corridors, cars or stores, any place where easy and quick exit is not available. Terror strikes rapidly with a choking sensation, a smothered feeling, pain in the chest, dizziness, numbness, tingling of the hands and feet, faintness, trembling, or fear of going crazy. It can be a combination of these. The reaction is "please, get me out of here!" Likewise, the fear of open spaces can occur in the supermarket, taking a walk, climbing a mountain, looking out of an airplane, anywhere in the open.

It is estimated that twenty-four million Americans experience a form of intense anxiety at some time in their lives, more women than men. Attacks can occur only once, once in a while or often. An attack can happen several times a day or every month or two. Some sufferers are afflicted only when under undue stress, others when there is no stress. A simple thing like sitting in the middle seat of an airplane can trigger it. There is no general pattern, but the reaction of disabling fear is similar in each case. The person never knows when it can happen, but is aware that it can at any time and learns to be prepared, avoiding situations that could bring it on. It is of the utmost importance to the victim to be sure of a quick and easy exit.

Panic attacks can be triggered by emotion, but often they become a pattern which repeats itself daily, and this is destructive to daily living. They are linked to depression and treating the depression sometimes cures the attacks. They are often coupled with the use of tranquilizers taken to relieve anxiety, which helps the anxiety but doesn't cure it. Alcohol can also touch off a panic attack, but the basic cause is the lack of physical activity. Many patients are cured by cutting out the drugs and exercising.

Panic attacks do not last long, merely minutes, but the effects can be lasting and color a person's life. The suffering is intense and usually prompts a person to seek medical care,as many take their symptoms literally. Fortunately, the episode, though powerful, soon subsides, but a minority of sufferers become confined to home, only venturing out with someone whom they can trust. The fear of an attack prevents many from going to the same location or repeating the same trip where one occurred. This can keep a businessman off planes and out of taxis and restaurants, gravely restricting his career.

There are people who fear sexual relations because of the anticipation of a panic attack, and avoid it completely. One woman refrained from a position of sexual interplay due to fear of being smothered, thus inducing hyperventilation.

The influence of panic attacks is far reaching. Some patients suffer from insomnia, others have the tendency to abuse alcohol and drugs. Caffeine is toxic to a few and trigger an attack. Another side effect is reactive depression, brought on by the patient's inability to lead a normal life.

The diagnosis of panic disorder can readily be missed. First and most important is the need for a thorough physical examination, as many cases are associated with abnormal heart rhythms or valve function which are easily treated. If a person has symptoms that relate to the heart and he is afraid he is having a heart attack, he goes to an internist or cardiologist, put through a series of tests and told there is nothing wrong. If headache or dizziness is his complaint, an internist is likely to prescribe medication, though such symptoms are often signs of panic. Persistent pain in the stomach sends him to a gastroenterologist and the tests and x-rays find nothing.

A psychiatrist is the most capable doctor to make the proper diagnosis, but it does not occur to these patients to go to one for physical symptoms, so panic disorder can be missed. Much time can be wasted before treatment is instituted. What is the treatment? The benzodiazepines are the drugs of choice in conjunction with psychotherapy and, while there are others, these are the safest, produce fewer side effects and act more quickly than the others, which take a week or two. They are alfrazolam, diazepam, ioprazepam, and chlorazepam. The anti-depressants are also effective in the treatment of panic disorders, both the monoamine oxidase inhibitors and the tricyclics, but all have side effects and take longer to act.

When a weak mitral valve is diagnosed, what is happening in the body? As the valve leaks, the blood pressure lowers slightly, lowering the oxygen content in the lungs. This frightens the brain because of inadequate nutrition, so the brain sends out an alarm and adrenalin is released to build up the blood pressure. This causes the patient to experience extreme fear which changes to panic. The patient reacts by sitting or lying down. This is the opposite of what is needed which is activity, for rest causes the carbon dioxide in the lungs to be exhausted and the oxygen to be increased. The resting body does not produce carbon dioxide so the lack of it produces the frightening symptoms of hyperventilation.

It is extremely important for the sufferers of panic attacks to know that help is available with a good chance of a return to a normal life. Here are examples. Lucy was happily married but had developed a fear of her mother-in-law who was a rather imperious woman. Every time she came to visit the young couple, Lucy had a panic attack and her shaking hand caused her to spill the tea on her mother-in-law's dress. Then she fainted. The entire family was upset and sent her to a doctor who taught her to breathe through one nostril and control the attack. Lucy finally overcame her panic attacks.

Jeanne was undergoing tremendous pressure taking care of her husband who was suffering from terminal cancer. One day, she heard that her daughter, her only child, was on drugs and had been divorced and her husband was awarded the custody of their two small boys. One night she was awakened at two in the morning with the devastating symptoms of hyperventilation. A visit to her doctor diagnosed panic attack and she was treated with drugs until the problems in her life were resolved. She is now free from attacks.

This is a complicated case, which shows that the treatment is focused almost entirely on proper diagnosis. Charlie went to a psychiatrist because of panic attacks which had begun to interfere with his work. The doctor sent him to a cardiologist who diagnosed mitral valve insufficiency and put him on a heart drug. However, the attacks persisted and he was referred back to the psychiatrist who determined that he also had endogenous depression, which was treated with drugs and psychotherapy. Charlie was cured of the attacks and the depression after a few months.

There are several causes of panic disorders but they are not well understood. One could be extreme anxiety to which some people are susceptible, and studies indicate that they tend to run in families, so genetic factors may be involved. Other studies show that they may be associated with the loss of a loved one.

4 PSYCHOSOMATIC COMPLAINTS

Psychosomatic medicine is not new. It was first described in ancient Chinese medical literature, and again in 500 B.C. by Hippocrates who claimed that doctors should know the "whole of things" in order to effect cures. Through the ages, this was reiterated by Socrates and the Roman physicians Soranus and Aurelianus, who practiced psychotherapy to relieve their patients' physical illnesses. The same philosophies prevailed in the Middle Ages, and it was well known early in the nineteenth century that strong emotional upheavals could result in disease, grief could cause illness, and one could, in fact, die of a "broken heart".

What happened to the whole person? There is a valid answer. Later in the nineteenth century Pasteur's work caused such excitement that doctors came to believe that there must be a special bacterium or organism responsible for every disease. Thus, physicians were trained to treat the disease and not the person. The concept of the whole person was shoved to the back burner, where it remained until recently. Today, the holistic approach to medicine is based on the treatment of the whole person, physically, mentally, and spiritually.

What does psychosomatic mean? The word comes from the Greek "psyche," mind, and "soma," body, or mind/body. It was coined by physicians in the 1900s, when they became aware that emotions and personality as well as physical factors can increase the intensity of an illness, and influence its onset, length and severity.

Somatopsychic, the reverse of psychosomatic, or body/mind, can also be used in describing an illness, for an illness that attacks the body also affects the mind. For example, acne, caused by an oversupply of oil secreted by skin pores, causes an emotional reaction. The person is ashamed of his or her appearance and the pores secrete more oil, worsening the acne.

What is a psychosomatic illness? It is when the nervous system influences the body in such a way that an illness follows. However, first it is essential that a thorough physical work-up be done to rule out physical disease. This will make sure that the mind is responsible. Every system in the body, such as eating, eliminating, coughing, kidney function, heart rate, and intestinal activity, can be influenced positively or adversely by the nervous system. These systems can also be affected by infection, an organism, an accident or the malfunction of an organ.

Psychosomatic medicine is divided into many areas and doctors vary in their understanding and application of the term. Many studies have been made but there is no consensus as to the cause, treatment or classification of these conditions, because the interrelationship between mind and body is poorly understood. Some consider psychosomatic medicine a specialty, and there are doctors who are experts in the field, but the scope is so broad and its delineation so blurred that the word is no longer adequate. Should the term be discarded? No, we are saddled with it because it is used universally, but it is unfortunate that its true meaning has become distorted. Doctors and laymen alike interpret it one way, that the mind is doing something to the body to make it sick. Some doctors use the word "functional" for psychosomatic which confuses the patient, When explained, however, some people become infuriated and others upset, and construe the diagnosis as mental illness, or worse, insanity. It is unfortunate that many people are still ashamed of mental illness.

So it is finally accepted that there are ties between the mind and the body that cause them to react upon one another. But how do they react? The mind can influence bodily functions to bring about or prevent disease, and the body, in illness, can manipulate the mind into a state of anxiety. We have also discovered that the mind can control functions of the autonomic nervous system like the heartbeat, digestion and the circulation. These had previously been considered solely under the command of a section of the brain called the hypothalamus.

Thus, the mind and the body are as one. A simple way to understand this concept is to visualize a human being as a set of scales, one side the body, the other the mind. Most of the time they balance. This is called "homeostasis". When one outweighs the other, illness occurs. If the body doesn't supply enough sugar to the brain, or if the brain doesn't supply the appropriate nervous impulses to the body, there is malfunction. The implication is that every system should be performing at its best all the time. Of course, this doesn't happen, but a slight dip of the scale one way or the other can occur and be easily put back in balance by a minor adjustment.

Our body systems check themselves, and compensate for an imbalance. If the kidneys work too fast, we go to the bathroom more often; if the nervous system is overactive, the body calls for more food or rest. When we don't pay attention, we might get sick. But unless the mind/body interchange is severely off, we adjust to imbalances.

We don't expect the automobile to run indefinitely without changing the oil, replacing the filters and spark plugs or having the brakes relined. We feed it with gas, just as we feed our bodies. Both use substances that cause wear and tear. The body, like a car, sometimes needs help in repairing itself. Our teeth need cleaning, diabetics replace the insulin they can no longer produce and we take calcium for strong bones as we grow older. These are not psychosomatic conditions, but normal wear and tear.

But if something breaks down, we hit the ceiling. "I never get sick! Am I falling apart?" Of course not, we just need a little repairing. But if we do become ill an emotional reaction takes over and muddies the waters. We can't have the flu without being upset about missing work, yet almost everybody gets the flu. We can't have arthritis without experiencing pain but we become very angry and feel guilty. This is the somatopsychic influence.

Our emotions are working overtime and emotional stress can bring on illness. the antithesis of the above. The loss of a loved one, or the sudden death of a dear friend, the break-up of a romance can disturb the body's homeostasis, thereby lowering the immune system, making it vulnerable to infection or disease, and the person comes down with a cold. Even a minor event can produce lowered immunity in some people. As we look back on life, we all have had this experience.

Listen to your body. It knows what it's doing even when you and the doctor may not. When you have an infection, you develop a fever and your white cells increase to fight it. This is the body's natural response. Given rest and time, the fever will come down. You may rush in to treat it, and end up fighting off these responses. If you are concerned, lie in a hot bath for fifteen minutes and the heat will fight the infection. Given a chance, your body will heal itself.

Here is an example of a condition that was emotionally induced. Marian's baby was a few months old when her daughter-in-law brought her first born home. Marian was teaching the young mother how to care for the baby, when she developed diarrhea and was afraid she might give it to the baby. Her physician found nothing wrong, and diagnosed a psychosomatic reaction to a difficult situation. Marian's son married while in college against his parents' wishes, and the arrival of the baby compounded the shock. Although reluctantly, she helped her daughter-in-law, for if she had not, she would have felt guilty. In that event, her condition may have become worse.

An emotional blast can turn inward or outward. We can take out our frustrations on family or become depressed. We could lose appetite, have diarrhea or a rapid heartbeat and hyperventilate. Or we could have none of these reactions. What began as anger at a minor illness developed into a chain of emotional and physical symptoms feeding into one another. And the doctor has to understand the whole history before he can make a correct diagnosis. He can't come into the middle of the picture, pick out a symptom and treat it successfully. So we have to cooperate with him and tell him everything.

We repeat, the correct diagnosis is important and must be made before starting treatment. The real challenge in psychosomatic medicine is in deciding how much of the illness comes from the mind and how much from the body. This requires a team approach by the internist and the psychiatrist and the first step is to know the patient as a whole human being. This method has been used for a relatively short time and has proved to be successful. The doctors, however, have difficulty working together because they speak different languages, the internist and the psychiatrist.

To make a diagnosis is difficult, and often the decision is psychosomatic. Mistakes are made by an inadequate work-up and jumping to conclusions. When a doctor can t make a diagnosis, he tells the patient: "It's your nerves", or "It's emotional", or "There's nothing wrong, it is in your head". And how does the patient feel? "I must be sick in the head." The cycle of mind/body-body/mind speeds up and the condition worsens. We see this with diseases of the skin, allergies, asthma, stomachache, diarrhea, loss of appetite, excessive appetite, ulcers, and serious conditions like ulcerative colitis.

This is an example of misdiagnosis. Christopher is depressed to the point of wanting to take his life because of prolonged diarrhea, loss of appetite and weight loss. Several specialists found nothing wrong, and psychiatrists treated his depression without success. After many drugs and therapies, he saw a doctor who ran more tests and found a parasitic infection which was promptly treated and cured. This illustrates the results of inadequate diagnosis. The patient's condition so depleted him that he could not live a normal life, and depression resulted. This demonstrates the effect of mind on body and vice versa, with mental and physical conditions worsening to an almost fatal conclusion.

So the doctor looks at both the body and the mind and determines how they are interacting. He gets to know his patient well, his personality and emotional makeup, and does a thorough physical work-up. In making the diagnosis, psychosomatic illness should be considered only after all physical causes have been ruled out.

Does everybody get psychosomatic illnesses? Everybody can, but not everyone does. The susceptibility is not tied to thinking and doing, but involved with personality emotional state, and life situation. It's like the chicken or the egg syndrome. Did something upset you, or did you come down with the flu first?

How can we keep this from happening to us? Accept your health, good or bad, with a positive attitude. Deal with yourself as you are, understand your body and know that at times it is not going to work right. Don't deny it if it doesn't, and it is equally important not to think you might have an illness. Set the goal of being as well as you can be. Believe that if you do develop an illness, you will get better. Accept it and deal with it positively. Don't fall victim to it. Sometimes you don't need help, sometimes you need a home remedy, sometimes you need a doctor.

Women are better at this than men. They have menstrual periods and bear babies. Men push being ill aside and deny they are sick, believing they will die if they admit it. It's foolish to expect to be in top shape all the time.

We can lose an arm or a leg and be a healthy person. We can have cancer and be a healthy person or a heart condition or diabetes. It's all in the attitude, and attitude can help avoid the emotional reaction that makes an illness last longer. Serious conditions can also be handled this way, by giving the body free rein to do its healing without interference from the psyche. This frame of mind can be cultivated, but it isn't easy.

How are psychosomatic/somatopsychic illnesses treated? Most are self-limiting and go away as quickly as they came, and if superimposed upon a physical illness, the treatment of that will eliminate the other. However, if the condition persists, psychotherapy is successful in every case.

How did the human race exist for four million years without doctors? Because the body is built to heal itself. Does the doctor mend a broken leg or cure a cold? No, we do. Most psychosomatic illnesses go away by themselves, which means they are self-limiting, like most illnesses. We don't know how, but we know that it happens.

II DEMYSTIFYING MENTAL ILLNESS
Mental Disorders From Chemical Problems of Thinking

PREAMBLE

Let us revisit the lake with two dams. It is oblong, with a dam at either end over which trickle the drops that represent persons with mental illnesses. Illnesses that arise from the thinking element trickle over one and those that arise from the doing element over the other, about one percent of each. The thinking disorders are chemical in origin and are called psychoses. The doing disorders are electrical in origin and called character disturbances. Seldom is there a mixture of the two in one person.

Over one dam go the psychotics with a chemical imbalance in the brain cells caused by a defective gene. They are schizophrenia, paranoia, schizoaffective illness, endogenous and involutional depression and manic depressive illness. In children and teenagers, there are also anorexia nervosa, bulimia and tic-like disorders.

Over the other dam go the character disturbances, also called behavioral or psychopathic disturbances. These come from electrical dysfunctions caused by inherited personality traits. They are the borderline, the psychopath, the career criminal and the sex offender.

There is a third group whose illnesses come from disease or injury to the brain. The location in the brain of the damage determines whether the patient becomes psychotic, psychopathic or mentally deficient.

How do the first two groups, the psychoses and the character disturbances differ? The psychoses are caused by an imbalance in protein metabolism, and can be corrected in most cases by drugs and with psychotherapy. Patients have periods of illness from ten to fifty percent of the time, during which they are more likely to be passive than aggressive. When they are ill they think and act irrationally, and seem to be not tied to reality. They appear introverted and wooden, and are awkward with other people. They show no feeling and interpret harmless remarks as criticism.

People don't understand these patients, which puts a strain on relationships.

Between bouts of illness patients function well, and may be business men, housewives, athletes or laborers. Unfortunately, psychotics are given inferior jobs, regardless of how intelligent they are. For example, Andrew, a bright young man became psychotic during military service. Fifteen years after discharge, his illness is under control. He writes well, reads extensively, but is only capable of menial work and moves from one job to another without a sense of purpose.

What about character disturbances? The man with a character disturbance, also called behavior disturbance, has an acting out problem that starts early in life. He may be aggressive or unobtrusive, and appear quite normal until his actions show him up. In either case, he is totally wrapped up in himself and insensitive to the needs of others. His amoral tendencies are not obvious when he is young, although he may kill the neighbor's cat or take money from the teacher's purse. Self-protection, however, keeps him from acts that could bring punishment. He is likely to alter his grades, cheat on exams, or steal from his mother and blame someone else, all of which he justifies by self-delusion.

The term "character disturbance" can be confusing as several adjectives are used in its connection, namely, criminal, psychopathic, behavioral. So let us stick to one and call these electrical problems of doing, psychopathic. Each personality has some degree, ranging from the petty thief to the hardened career criminal.

The spectrum of degree is wide. On the lower end is the habitual thief, motivated by easy money and adept at evading the law. He is loyal only to his group. At the other end is the full-fledged psychopath, loyal to nothing and no one but himself. Both are clever at escaping detection, but the difference between them is motivation. The thief is motivated by greed or drugs, the psychopath for the satisfaction of breaking the law and getting away with it.

A newspaper headline reads: "Ex-mental patient pushes blind man under subway train". The culprit is arrested and sent to a hospital for observation, where he is declared a "mental patient". This label suggests a psychosis but there is nothing wrong with his thinking but something very wrong with his behavior. Why did he do this dastardly act? Because he is an extreme doing person and got drunk, and the alcohol released all inhibitions. He is psychopathic and a dangerous criminal. The psychotic hallucinates, says strange things, and withdraws into his world, but he doesn't harm anyone. He is the mental patient, not the other one. That is the difference between the psychopath and the psychotic. Most people, psychotics included, have a built-in factor called the "restraint mechanism" that prevents them from hurting people and committing crimes. It is part of

the "conscience". The psychopath lacks this. He has no conscience.
Today, modern technology offers an accurate diagnosis of these illnesses by brain imaging. This was not possible before the first quarter of the twentieth century. Before that time, psychiatrists and scientists knew that mental illness is caused by disturbance and/or malfunction of activity in the brain. Dr. Emil Kraepelin brought several brain scientists to Munich University to search for the causes of these disturbances. Although they could not research the brain except at autopsy, they were able to lay the groundwork for the study of modern neuroscience.

These scientists identified schizophrenia and manic depressive illness, and isolated the pathology for Alzheimer's disease. In view of subsequent research findings, it is clear that they were on the right track, but they lacked the sensitive tools that are available today. Subsequently, the 1980s brought brain imaging, with the ability to identify the anatomic, metabolic, and neuro-chemical substrates of the psychoses.

The symptoms of mental illness may be classified into two groups, those that are an exaggeration of normal thoughts and those that are a loss, sometimes called positive and negative symptoms. The positive are due to hyperactivity (high) and the negative to hypoactivity (low). This is over simplification, but serves as a framework for research. Positive symptoms are distortions or exaggerations of normal thinking. They may be auditory hallucinations in schizophrenics and manic depressives, when the patient hears voices telling him unrealities. Presumably, these aberrations occur spontaneously within the brain without outside stimuli, and consist of delusions and false inferences about people and things. A man might claim he is Jesus Christ or Napoleon Bonaparte. Negative symptoms are the absence or lessening of ordinary behavior patterns that should be present, such as the display of emotion, lack of emotion, speechlessness, inability to concentrate.

Both positive and negative symptoms are difficult to associate with brain imaging, as we have not yet reached the point where it allows us to identify psychotic and psychopathic pictures of the brain. This is because electro-chemical changes are taking place inside or between the nerve cells and are too tiny to be photographed by today's technology, so they are invisible.

However, we can see chemical changes on a PET scan (positron emission tomography) taken after injection of the patient with a radioactive substance. This shows large areas of certain chemicals present in some schizophrenics that would not be there normally. This methodology is experimental and seldom used, although the radioactive substance rapidly decomposes and does no harm, but there is some risk, so we are a long way from the accurate diagnosis of psychotic and psychopathic disorders by this means. It is helpful, however, in visualizing chemicals in the brain but is not necessary in order to make a diagnosis, which can be done by psychiatric examination.

The encephalograph can record electrical activity in the brain and take pictures of malfunction responsible for aberrant behavior, but cannot diagnose psychiatric conditions. The pictures do not show that type of pathology. They do show seizures and areas of dead brain from injury and disease, and weak transmission, such as by a tumor. But a neurosis looks just like a psychosis on the encephalogram, and we can't tell a psychopath from a schizophrenic.

So what does this new brain imaging do? It makes it possible for neuroscientists to study the anatomy of the brain, to map changes in metabolic activity and to measure the chemical action of neurotransmitters in the brain. Thus they can compare normal brain function to that of persons with mental illness.

Until the development of these diagnostic tools, animal research was the only method of studying brain function, but there are no animals that correspond to the knowledgeable and emotional experiences of human beings. Only humans can create abstractly and communicate in a technical verbal manner. Animals cannot do this, so most of the research must be done on human beings while they are alive. We learn much from post mortem brains but this does not offer the opportunity to witness the brain in action, which is difficult and complex, as brain systems and circuits are not independent of one another. They all interact. However, it's an exciting challenge to be able to identify the mechanisms that cause abnormal thinking and behavior.

Modern technology developed CT, (computerized tomography), in the 1970s. CT measures structures and for the first time permits the brain to be studied in its living state. CT scans are useful for a variety of conditions, and are now used extensively as a differential diagnostic tool, such as a tumor from a blood clot and a fractured skull from an ordinary headache.

Then came MRI (magnetic resonance imaging), which supplements what CT cannot do, so the anatomy of the brain and structural abnormalities can be studied with these two machines. CT images in a transverse plane, MRI images in all planes, and doesn't use ionizing radiation as CT must. MRI delivers a three dimensional picture of the brain and can be applied to four areas. It is a radio wave picture in three dimensions. The CT scan is an x-ray picture in three dimensions. We are not measuring function with these machines, just taking a different kind of picture. The difference is that x-ray penetrates substance, whereas a radio wave bounces off substance, producing a reflective picture with the MRI of cells that contain water. Thus the picture is really of hydrogen, (H_2O), as the water content of bodily substances reflect radio waves, giving a picture of the hydrogen content.

So the MRI can picture everything but bone, x-ray nothing but bone. The MRI takes a picture nicely of soft tissue which is difficult to do otherwise, but it cannot take one of hard tissue, such as bone. So MRI is far more efficient than CT because of precision in visualization, and is more adaptable to the study of parts of the brain not visible on CT. So today, we know a great deal about normal and abnormal brain activity, but this is not visible on the CT or MRI scans, but can be identified through psychiatric examination and by talking with the patient.

Scientists now can see the brain at work by means of SPECT and RCBF, (single photon emission computed tomography and regional cerebral blood flow). SPECT is a scanning technique using the computerized tomographic reconstruction method along with single photons emitted through a tracer administered to the patient. It can produce images in multiple planes. SPECT is used in the study of Alzheimer's disease and depression. RCBF shows hypoperfusion (low blood flow) in a certain part of the brain which appears to be an earmark of Alzheimer's disease. In depression, blood flow patterns are different, making the differential diagnoses of dementia and depression more readily made.

SPECT is an important adjunct to PET (positron emission tomography), which is a more sophisticated and more expensive technique. SPECT's resolution, however, is not equal to that achieved by PET, but suffices in many applications. PET is the best brain imaging technique in sensitivity and flexibility, and has two applications, the assessment of metabolic activity and the measurement of neurotransmitter function.

Currently, investigators are using PET to study neuroses and other less serious mental illnesses. An area of hyperactivity has been pinpointed, indicating a susceptibility to lactate induced panic attacks. Because this particular area in the brain records memory, the presumption is that the attack may be triggered by a former experience that produced severe anxiety. Increased activity has been noted also in patients with compulsive obsessive disorder.

Studies are ongoing of neurotransmitter systems in the brain, dopamine receptors and the mechanism of neuroleptic drugs, which will lead to a clearer understanding of schizophrenia. Scientists can now study the brain directly, which increases the understanding of side effects, such as tardive dyskinesia, which can result from drugs in the management of psychoses. It is also useful in the development of more effective pharmacologic agents.

Brain imaging offers many techniques to investigators in their attempts to unravel the mysteries of mental illness. It allows the ability to "see the brain" structurally and functionally. It is useful in differential diagnosis and gives additional information about the pathology of mental illnesses and metabolic abnormalities. It promises the mapping of cerebral function in normal persons and thereby increases our knowledge of normal and abnormal brain function.

This is a case history that illustrates how little we know about mental function that the most sophisticated machines cannot delineate. A sixty year old man complained of a variety of minor symptoms, fleeting headaches, fatigue and so forth. He was a heavy smoker and x-ray diagnosed cancer of the lung. Suspecting metastases, an electroencephalogram was done and found normal, but x-ray showed the brain riddled with cancer, subsequently confirmed by autopsy.

This patient was not neurotic or psychotic, and his severely diseased brain caused only minor symptoms. Three doctors visited him in his hospital room, and to test the patient's thinking, one asked him to name a few simple objects which he did quickly and accurately, then asked how many men were in the room. He replied: "Four". He was asked "Four?", "Yes", he said. "I'm a man too". This man's thinking was perfectly clear despite his severely damaged brain.

The lesson learned from this is that we cannot always equate physical damage with emotional damage. And the finest machines cannot tell us as much as observations by the physician, the psychologist, the nurses or the family, who are in a position to observe the most closely. So far, there is no way in which thinking and behavior can be tested.

This explanation of brain imaging may be confusing to some readers, so let us recapitulate. These new methods provide better pictures of all parts of the body more clearly and safely than the old standbys, the x-ray and the encephalograph. The x-ray shoots particles through substances and takes a picture that penetrates the object being studied, such as the chest. The particles go through muscle and other tissues and are stopped by bone. As they strike the x-ray film, they make a picture of everything except the bone. This creates black areas where there is soft tissue, saturating the film. It is like taking a picture in reverse, where there is hard tissue, such as bone, nothing shows. Denser tissues are not well defined because they don't stop the x-ray well, but show up vaguely. The result is a black field with a white skeleton on it.

The encephalograph has been in use as long as the x-ray and is similar to the electrocardiogram, which takes recordings of the activity of the heart. The encephalograph records electrical activity in the brain or the lack thereof, and takes pictures of malfunction, but shows nothing else.

The CT scan uses the same x-ray particles and shoots them through the body, taking pictures in series at various angles. Then the machine computerizes them to give a three dimensional picture. So the CT scan delivers a better picture than the x-ray of tissues.

The MRI is a picture taken by radio waves that bounce off and are reflected back by the tissues and then by hydrogen (water) atoms. Radio waves are less damaging to tissues and less apt to cause illness than x-ray waves, and can be bounced off any kind of atom. Hydrogen is used the most as there is little hydrogen in bone, and the picture of soft tissues full of water is superior because the reflected particles are bouncing off it. This makes the picture a reflective wave, rather than a wave of particles going through something.

The SECT is also a radio wave, but is picked up by a scintillation crystal rather than an x-ray film. The picture is computerized so it has three dimensional effects of multiple images at different angles. Again, the picture by radio waves is safer than by x-ray, and the scintillation crystal is more accurate and can pick up smaller amounts of reflected energy. As an adjunct to SPECT and for the diagnosis of diseases such as Alzheimer's, regional cerebral blood flow, or RCBF, is used.

The fifth and most sophisticated and expensive machine is the PET which requires injection of the patient with a radioactive substance. This technique is experimental and poses some risk. It highlights chemicals in the brain that are abnormal, which is helpful in the diagnosis of psychoses.

With these advances, we can take pictures of the body's soft and hard tissues, nerves and bones, in fact, all the substances in the body. This gives us diagnostic opportunities we never had before. However, the information offered does not negate the knowledge, experience and skill of the physician who interprets them and links them to the symptoms, the personality and the medical history of the patient in an organized fashion. It is unfortunate when a doctor depends more on a battery of tests than on his ability and experience to make a diagnosis. After having made it, he can employ the tests to clinch it. Thus the diagnosis can be confirmed.

5 THE PSYCHOSES

The psychoses are mental disorders of thinking and mood stemming from chemical problems in the brain. They are caused by a defect of protein metabolism in the brain. We know what thinking is, but what is mood? Mood is your state of mind, whether you are in high or low spirits or somewhere in between. A low mood that persists may mean depression, just as high spirits that persist may mean mania.

We will explain fully later.

What is metabolism? It is the means by which food passes from the intestines into the blood stream where it is transformed into substances to be used by the body. These substances are carried by blood vessels into the brain through the blood-brain barrier, where they are altered and used by the brain cells. What is the blood-brain barrier? It is a mechanism that isolates the brain from the rest of the body and screens materials that enter it. There is a special structure located in the capillaries that surround the cells in the brain that materials have to pass through to enter the cell membranes, rather than through clefts in the capillaries as they do throughout the rest of the body.

Thus, the brain is protected, and receives only some of the substances that are taken into the body. For example, large proteins are not permitted entry, nicotine and alcohol are allowed in. So the brain is protected by the blood-brain barrier from many of the harmful substances that otherwise could enter, however, many dangerous and even lethal chemicals can penetrate the barrier.

Protein metabolism starts in the liver and is completed in the brain where both provide nourishment for the cells and for their products, i.e., neurotransmitters. We have long known that the human body needs protein in the form of meat, fish, dairy products, nuts and legumes. Studies show that a thinking person needs more protein than a doing person because his brain uses more amino acids to operate his mental function.

A psychosis occurs when the cells don't produce neurotransmitters correctly. Why, we don't know. Eventually, research will reveal the answers. What is the difference between a person with a psychosis and one without? Both may experience depression or elation caused by an imbalance of neurotransmitters, but a mild imbalance straightens itself out automatically, while the person with a psychosis maintains a permanent or long lasting imbalance. Dietary protein does not change the brain proteins, but if the patient is treated by a drug which can correct it, the brain balance will be normalized and usually stabilized for good.

All of us have an imbalance in mood at times, one day up, the next down. The balance quickly returns as the body converts protein into amino acids and alters them to a form the brain can use. These amino acids are the chemicals serotonin, noradrenalin, dopamine and others we don't yet understand. This occasional imbalance is reflected in our dreams and quickly rights itself on awakening. Sometimes dreams spill over into daytime thinking, showing up in unrealistic thoughts, even in fantasies that are inane.

At the cellular level, metabolism obeys the laws of chemical thermodynamics, which is the science of energy changes, and is governed by the role of matter and energy in physical and chemical systems. The law that is important to us is that all chemical reactions proceed in the direction that releases energy. As we explained, this is the basic function of metabolism, and the substances produced are carried by blood vessels into the brain through the blood-brain barrier, where they are altered and metabolized by the brain cells which utilize them. Every cell in the brain metabolizes sugar for energy and protein for neurotransmitters or other transmitting substances of internal chemical function.

Then the body can metabolize any one of the three substances, protein, carbohydrate or fat, and each of these can be metabolized and altered from one to the other and again metabolized. Thus, sugar can be made out of fat and protein out of carbohydrate.

There are certain essential amino acids and fatty acids that cannot be produced from other forms. These have to be taken into the body as foods, because the body cannot produce them chemically from other molecules. So these essential substances have to be eaten in order to make it possible for the body to produce the others from a simple form.

Back to the blood-brain barrier. Some drugs are allowed entry through it, such as caffeine, marijuana, nicotine, alcohol and hard drugs, i.e., codeine, heroin, cocaine, LSD and so forth. All of these are potentially harmful. They are mind altering drugs and give the individual a pleasant sensation of euphoria. Experiments on small animals demonstrate that when rats and mice are given one of them they will refuse food and take only the drug. Eventually, they die. If a person begins to take one regularly, he may soon find that he cannot stop. He is addicted. So it is up to the individual to decide which and how much of the harmful drugs he or she will allow to enter the brain and for how long.

Most substances in the form of foods that are good for us cannot enter the brain. This means that we cannot supply our brain cells directly so they can manufacture neuro-transmitters that are essential to health. But in nourishing the body, we know that the brain will select those molecules that it must have to produce neurotransmitters. However, we cannot overload the diet with foods that produce them and expect the brain to let them in.

Therefore, our diet should consist of a mix of protein, fat and carbohydrates to adequately supply the body's needs. An overabundant diet is not going to make the brain cells work faster and this only results in putting on weight we don't need. So what we put into our bodies as food will permit the brain to produce the energy it requires in order to function at the optimal. If the body is not properly nourished, it can struggle along, but not at the level of which it is capable.

This is the extent of our knowledge today. Ultimately, treatment will be the ability to introduce into the brain metabolic function by genetic engineering. One method may be the transplantation of new cells into the brain so the neurotransmitters can reproduce naturally. At this point we cannot change the cells, but we can change their metabolic rate by medication. It is not perfect but the best we can do now. Eventually, research will reveal the answers.

NEW HOPE THROUGH GENETIC ENGINEERING
Genetic manipulation and gene targeting began ten years ago when genetic engineering was first discovered and is now being put into practice

on animals and soon on human beings. Expectations are high in the search for ways to treat or eliminate genetic diseases, of which there are about three thousand known, each caused by the failure or abnormality of a single gene. For most there is no treatment or a treatment that is merely marginal.

The genes for Downs syndrome, sickle cell anemia, Huntington's disease, manic depressive disorder, muscular dystrophy, cystic fibrosis and familial hypercholesteremia have been identified, but not yet are those responsible for addictive diseases, such as alcoholism, cocaine and heroin abuse, schizophrenia, major depression and the behavioral disturbances. How do we go about this? The first step is to find the gene responsible, then comes targeting it or finding the position on that gene.

It is like looking for a needle in a haystack but will lead to effective therapy for these incurable conditions.

Replacing bad genes with good ones is not easy, and scientists are having difficulty transplanting genes to a specific location on a chromosome, as they sometimes are inclined to attach themselves to unpredictable places. This could be crucial, as it might inactivate an important gene or activate an oncogene to turn into cancer. Nevertheless, scientists have developed a new technique whereby it will be possible to transplant, substitute or modify almost any gene, thus replacing bad ones with good copies. Up until now, new genes enter the hereditary material at unpredictable places rarely hitting the target. The new technique causes the transplanted gene to settle in the right place.

Genes from other animals and from human beings are being transplanted into mice, goats, sheep and pigs, resulting in species called "transgenic", to be used in research. These transgenic animals serve as models for which scientists can develop therapies or eradicate their incidence in future generations. Thousands of transgenic mice have been created by inserting into a mouse-fertilized embryo the DNA from another animal, then placing the embryo into a female mouse. The fetuses develop normally and are born carrying the foreign gene and their offspring also carry the gene.

Thus, researchers have been able to produce milk from mice that contains a human blood substance called "tissue plasminogen activator" or TPA. This is used on patients immediately after a heart attack to dissolve blood clots and has saved the lives of many. It is estimated that one hundred transgenic goats can produce enough milk to supply TPA for every heart attack that occurs each year in the United States.

This technique has moved into the industrial arena to produce valuable drugs and other substances. Specialists in agriculture hope to improve the quality of meat and create resistance to disease in animals. They have already created transgenic cows, pigs, sheep, and goats, as well as fish that grow faster and larger. There appears to be no limit to the possibilities. This transgenic method is less than a decade old, and is being worked on in schools of medicine and research institutes around the world. Some of the diseases being studied are cancer, infertility, AIDS and human growth hormone. The Du Pont Company has developed transgenic mice that carry the gene for human breast cancer which should advance the testing of anti-cancer drugs. We have found the gene for manic depressive illness, and it should not be long before those for other psychoses are identified.

A tremendous growth in the field of transgenesis is taking place. Scientific experts foresee that resistance to disease in the form of an augmented immune response may be an important by-product of these studies, which can be applied to humans as well as agricultural animals. The present and future looks bright for genetic engineering.

6 SCHIZOPHRENIA, THE MOST COMMON PSYCHOSIS

Until the 1920s, "dementia praecox" was the term for an illness of the mind which has afflicted mankind throughout history. In the middle 1800s, the term was used by Morel, but the condition was not described in detail until 1898 by Kraepelin. Actually, it was a trash bin for mental illnesses, called insanity. That word is no longer used today. At that time, mental illnesses were not identified or understood, much less helped or cured. As doctors became able to recognize mental disorders, in 1911 Blueler named the most common "schizophrenia", from the Greek "schizein", to split, and "phren", mind, which he felt was a more fitting description. Schizophrenia does not mean a split personality, but a split in thinking between the real and the unreal worlds the schizophrenic alternately inhabits. So schizophrenia is a thinking illness.

Schizophrenia first appears in adolescence or young adulthood, and is a leading cause of teenage suicide. It is a psychosis, a chemical problem in the brain cells, and an inherited defect in protein metabolism. A recent study by the National Institute for Mental Health reveals that one percent of the population of the United States are afflicted. Ten percent of these are from families that carry the gene. Mild cases are easily controlled and some are cured, the more serious are amenable to treatment, but are disabled to some extent. The most severe are resistant to treatment and are totally disabled.

The tragedy of this illness is the devastating effect it has on brilliant minds. Jonathan, a bright young stockbroker became schizophrenic and could no longer handle the work. He is now a courier. Philip was Phi Beta Kappa in a top university and went to medical school. There, he was tempted to play with drugs and became addicted, developed schizophrenic symptoms and attempted suicide. The drugs increased his tendency to schizophrenia and his medical career was destroyed.

Marjorie, a brilliant student, became schizophrenic and suicidal after the birth of her first baby. She was unable to care for her child, broke up her marriage and went to live with her parents. She hates her surroundings and hates herself. She has become isolated from life.

Caroline, a young society girl, was talented and beautiful. While studying for the stage, she began to have delusions that she was a famous actress and expected her associates to treat her as such. She lived in an unreal world. Schizophrenia was diagnosed and she committed suicide a few years later. The above examples show schizophrenia in its most severe form.

How does it show up? It can happen slowly and insidiously or suddenly. A young person may make a remark with no basis in fact, such as "the television tells me what to do" or may become uninterested in life and complain about imaginary problems. Sometimes the person suddenly announces he is someone else, like General MacArthur or Florence Nightingale. The thinking is not clear, something has gone wrong.

As only thinking people have this disorder, the schizophrenic is extremely sensitive and perceptive, which goes along with the thinking element. The mix-up in brain chemistry accentuates thinking to the point where a patient sees and hears strange things. He seems to be living in a world apart, and the description of what he is experiencing is, to us, disturbing and even frightening. Schizophrenics are seldom dangerous, however, and have many moments of clear thinking, and can converse about current events as intelligently as anyone.

Many great inventors, artists, religious leaders, poets, scientists, chemists and other innovators are deeply intellectual and all are extreme thinkers. They are often considered strange, a little crazy or daffy, and some are schizophrenic. Two of the greatest chess players, Brophy in the last century and Bobby Fisher in this one, became schizophrenic after beating every opponent. Neither was able to play again. It is a mystery why these men broke down, but they were certainly committed to a single line of reasoning.

History shows that exceptionally creative people have a high incidence of schizophrenia. Schizophrenics are intense, and mental stimulation creates high mood which arouses paranoia, during which period their best work is done. This increases tension and one feeds on the other until a chemical imbalance occurs causing emotional distress. This drains them of the high mood, then they crash into depression. The famous artist Vincent Van Gogh, was one of these. During his high mood, he created a new work of art almost every day for a month.

It is important to understand these people and accept their violent mood swings, the highs of overactivity, the lows of depression, the suspicion, the dullness. They are difficult to deal with in any of their moods. But schizophrenics are invaluable to progress, and may contribute more than stable citizens. The world needs them.

There are five types of schizophrenia, catatonic, disorganized, paranoid, undifferentiated and residual. These are not cut and dried categories and a person can have one type or a combination of two or more.

The *catatonic type* is a common form and the most dramatic. Catatonia is characterized by almost complete immobility and total silence, alternating with frantic overactivity.

When in catatonic excitement, the patient runs about frantically, jumping over the furniture, and banging on the walls. Hyperactivity is so extreme that he may collapse in exhaustion.

Some cases never enter the mobile stage, but remain in a catatonic stupor and require total care. They adopt bizarre statue-like positions which they hold for hours and are totally out of touch with reality. They have delusions and hallucinations of a vivid nature. Negativism is a feature of catatonic schizophrenia. When asked to do something, the patient does the opposite, if told to stand up, he lies down. If asked to open his mouth, he clamps it shut. Echoing is another symptom. When asked "How do you feel today?" He repeats the question. He seems unable to put meaning into the spoken word. Miraculously, some of these patients recover spontaneously or with treatment by drugs. Here is an example.

Madeline, twenty-two, was brooding over life, did not feel well, but couldn't say why. She wanted to be with people and sought them out, only to break off a conversation suddenly and move away. One morning, her mother found her motionless, with her legs twisted into awkward positions. She wouldn't talk or move, and acted as though she didn't see or hear. Her face was expressionless, and her terrified mother called the ambulance. After a few days, she spoke but only repeated questions. She made no attempt to care for herself. Electric shock treatment resulted in total remission.

The *disorganized type* (until recently, called "hebephrenic") is the most difficult to diagnose. The onset is insidious and deterioration is rapid. The patient is extremely disorganized, his thinking is irrational, his conversation incoherent, silly and illogical, and his emotional reactions are unnatural and erratic. Hallucinations are common but not unpleasant to him, and he may experience delusions of grandeur.

These patients are obsessed with the idea that weird changes are taking place in their bodies. They claim that the heart is moving about, the kidneys are lodged in their legs or the brain has melted. They are lethargic, indifferent, and detached, and talk unintelligible gibberish, referred to as "word salad". As the condition worsens, they neglect personal hygiene and require constant care. Here is an example.

Alexandra, seventeen, did well in school until recently when she became apprehensive about her homework. She was overly concerned about approaching examinations, which she claimed she was unprepared for. She kept repeating, "Should I go to school? Will I pass?" and no amount of reassurance registered. One night she threatened to kill herself and was hospitalized.

She was confused and hyperactive. She resisted offers of help, and was unable to answer questions except with whimpering sounds. She was agitated and kept picking at her face, screaming or staring out the window. She refused to eat and laughed illogically, babbling incoherently. She explained that the Russians had put her in a concentration camp. Treatment with psychotherapy and drugs resulted in slow improvement.

Paranoid schizophrenia can appear at any time of life, at puberty, during adolescence, in middle age. Late in life cases are difficult to distinguish from the psychosis, paranoia. These patients make irrational statements, their paranoid element is accentuated, their thinking is unreal, and they relate everything to themselves. A patient may have delusions of grandeur and imagine he is Napoleon, a famous actor or statesman. A female patient believes she is a renowned ballerina, celebrity or artist. They hear voices and experience hallucinations involving all the senses. They demonstrate the attributes of the universal personality trait of paranoia in an accentuated form. They believe they are being followed and accused of breaking the law, and may suspect a man coming to repair a telephone of being a spy. All events are projected from self back on to self, in the belief that hidden forces are working against them. These are all signs of excessive paranoia.

As the illness progresses, delusions become pleasant, and the patient welcomes them and builds on his grandiose ideas. Unreality occupies his mind. Unbelievably, however, emotions and speech remain coherent and comprehensible. The difference between the disorganized type and this type is more rapid disintegration. Here is an example.

Albert had been normal until he went to college and began to experience difficulty keeping up with the work. It became so hard for him, he dropped out and took a job as a salesman. This did not last long because he was convinced people thought he was homosexual. When he heard the word, home, he misconstrued it as "homo", and fair, as "fairy". When he saw people talking on the street, he was sure it was about him. He became more and more disturbed and claimed he heard people saying things about him. This upset him so much he quit his job. Soon, he became restless and disoriented and was hospitalized with a diagnosis of paranoid schizophrenia. Treatment with medication and psychotherapy improved him and he returned to a new job.

Twenty-five years ago, when a teenager, Pamela was diagnosed as having intractable paranoid schizophrenia. She suffers from terrible delusions and paranoid thoughts, voices that torment her and hallucinations that frighten her. She has been under the care of one of the authors from the beginning of her illness and has been admitted to the state hospital twenty times, where her condition is stabilized with medication. She then returns to her husband and children and resumes her position as a personal secretary. However, she soon stops her medication and regresses into illness again.

She is a good worker and well liked at the office as well as in social life, but the problem is that she wants to stop her medication. You would think she would be delighted to be free of the horrible symptoms she suffers every time she stops it and usually ends up in the hospital again. When she takes it, she leads a normal life with her family and should realize what it means when she neglects to take the pills. But no, she is incapable of admitting that her thinking is unrealistic and continues the cycle over and over.

The last time the doctor saw her he went over her chart, showing how the medication helps her. Why does she keep wanting to stop it? Every time she does, she goes back to the hospital where she is medicated and sent home again. Is that what she wants for the rest of her life? This is typical of most psychotic patients.

The *undifferentiated type* is, as its name implies, the type of schizophrenia that does not meet the criteria of the catatonic, paranoid or disorganized types. These patients have prominent delusions and hallucinations. They speak incoherently and act in a disorganized manner, and some may exhibit several of the symptoms of the other types. Here is an example.

Jonathan became confused as a high school student, rambling about incomprehensible ideas to his friends. His marks steadily worsened and his teachers noticed that his homework was never completed. After an extensive medical workup, he was referred to a psychiatrist who sent him to a well known mental health in-patient facility. After several years of multiple therapies, he was less severely ill and returned home but was unable to return to school. He tried several entrance level jobs but could not concentrate well enough to continue, so he applied for social security disability. He now lives on this income in a half-way home and spends his days in social activity as directed by his day care center.

The *residual type* of schizophrenia is not dramatic. It starts in a subtle, insidious way before puberty usually unnoticed until the child gradually withdraws into a constricted life. He refuses to go to school or leave the house, is inactive and, as he grows older stays at home. His thinking is logical but limited. He has few ideas, talks only about concrete events and cannot think in the abstract. There are no delusions or hallucinations. He is simply unable to cope with the demands of life. These unfortunate people become burdens to their families, and if they are rejected they end up as hoboes, Bowery bums, prostitutes, or bag ladies. Here is an example.

Susannah became more and more dependent on her younger sister. Her father died when she was a child and her mother during her teens. She had several jobs, but was fired or left each one, saying: "It's too much for me". She led a restricted life and refused invitations to go out. Finally, it was agreed that she would do the housework and her sister would go to work, but even that became "too much for me". She sat for hours, seemed to have no interest and gradually relinquished her household duties. Her sister sought help and she was institutionalized. This is the schizophrenic of the residual type.

Women who have residual schizophrenia are often prostitutes and many have been examined psychiatrically. A large percentage are psychopathic, have poor morality and are just out for the money. They take the easy way to earn a living and don't give a whit about passing along disease.

If such a person has a high degree of paranoia, she will show a tendency to isolate herself and exhibit a fear of being approached sexually. These women are sexually very active or very inactive; there is no middle ground. Yet the desire to have a baby is strong and often they will become involved sexually, fall in love, get married and have children. Then they are extremely hostile to their mates, scold the children for the slightest misdoing and blame them for things that go wrong.

These marriages usually end in divorce.

There are rare cases of mental illness that resemble schizophrenia and with similar symptoms, but come from an organic disease. These show up at any time from childhood to old age and are difficult to diagnose and treat. For example, an epileptic woman has visual and auditory hallucinations when she has a seizure. She sees God, dressed in white, standing on a hill, and talks to Him and He talks to her. She is not schizophrenic, her hallucinations follow a seizure, which creates an electrical discharge, disrupting brain chemistry.

An extremely introverted personality is sometimes suspected of being schizophrenic. If a person is shy and withdrawn, chooses solitary work and prefers to be alone, people question his mental health. Many people such as this lead productive and exemplary lives, but if they are schizophrenic, more dramatic symptoms will appear in time.

Some patients have more than one illness and are the most difficult to treat because one intensifies the other. The epileptic who also has schizophrenia is treated less effectively because the drugs for schizophrenia make the epilepsy worse. The schizophrenic, who is also alcoholic or has an addictive problem, is difficult to treat because alcohol makes him feel better but makes the schizophrenia worse. And as the drugs he takes for schizophrenia give unpleasant side effects, he turns to alcohol to feel better. Then the effects of alcohol cause him to stop his medication for schizophrenia and he becomes more psychotic. A catch twenty-two.

The worst cases and the most unhappy are the mentally retarded who have schizophrenia and are also addicted to alcohol or drugs. It is impossible to treat them because they don't have the intellectual capacity and don't fit into group therapy or Alcoholics Anonymous, and psychotherapy is of little value. Placement in controlled settings is the only recourse.

Of all mental disorders, schizophrenia is the most difficult about which to predict a prognosis because of its complexities. Some patients recover with treatment in a few hours or days, others remain ill for the rest of their lives, and some undergo a slow, downhill course.

Is a schizophrenic's brain different? Yes, it is not the same as a normal brain. Doctors have found that it is distorted or shrunken, however, the main difference is in the chemistry within the cells themselves. There is a surplus of dopamine, the by-product of the metabolism of the amino acid tryptophan, with which the brain is flooded. This super-abundance causes the symptoms, which range from the overactivity of the disorganized to the underactivity of the catatonic, from thinking too much to thinking too little, from pleasant delusions to agonizing hallucinations. This imbalance produces a variety of visual, auditory and other sensations. Each case is unpredictable, due to the different ways the brain reacts to the surplus of dopamine. There is a theory now being researched, that schizophrenics have too many dopamine receptors as well.

What about sexuality? How do schizophrenics handle it? The male has difficulty with sex. He is capable but unable psychologically. One patient could have intercourse only with prostitutes, but would masturbate first because he couldn't reach orgasm; his mind was blocked when another person was involved. So he visits a prostitute who may be a residual schizophrenic and although he cannot function and neither has an orgasm, they spend a day or night together and he pays the fee of five hundred or more dollars. They play with one another and may bathe together and touch and fondle one another, but never have sexual intercourse. They may attempt it, but it never happens. He feels, however, that the experience is worth the money because he has sexual contact and considers this a sexual experience.

The homosexual schizophrenic also has a problem with sex, and although he thinks a great deal about it and is physically capable, he is unable to perform. The schizophrenic seems to have an irreparable blockage in regard to sex between the thinking and the doing.

Who becomes schizophrenic? We know that schizophrenia is hereditary, but is that all? In spite of an enormous amount of research, scientists have yet to pinpoint the cause. Some claim it is a recessive gene, and most experts go along with this. Some suggest a metabolic origin, the environment or a predisposition, and others the endocrine glands. Recent research points to a virus, another is the time of the year the patient is born. So far, these theories are unsubstantiated.

We do know that at birth, or at puberty, something happens to the chemistry in the brain cells resulting in an imbalance and a breakdown in the chemical connections within the cells. We also know that if both parents have schizophrenia, their child has two out of three chances of having it. If neither is schizophrenic, there is still a risk if the thinking quality of the parents is excessive. Families with a history of the disease have most of the schizophrenic children. It is uncommon in those with no history.

Before 1954, there was no treatment for schizophrenia. Patients went into state hospitals where they lived as comfortably as possible. They performed minor jobs like washing dishes and scrubbing floors, because their disordered thinking precluded anything more complicated. In 1944, thorazine, the first of the phenothiazine drugs, was discovered. Patients were treated with it and sent home because of the great improvement.

Now, there are three forms of treatment, physical therapy, chemotherapy and psychotherapy. Physical therapy is mainly electric shock, used in severe cases to return a patient to a stable condition. However, the benefits are short lived. Chemotherapy, or treatment by drugs, vitamins and minerals, with psychotherapy is the treatment of choice. This reduces the amount of dopamine in the brain and/or blocks the dopamine receptors, and is effective in most cases.

Thorazine and other similar drugs are neuroleptics, which are anticonfusional and anti-psychotic medications. They stabilize the chemical imbalance and allow the thinking to return to normal. A few cases don't respond to them, and are treated with limited success by psychotherapy and electric shock along with other drugs, such as the anti-depressants and vitamins and minerals. Even though patients don't hallucinate or have delusions, medications don't restore thinking completely to normal, so patients are often unable to adjust to society and cannot cope adequately with life. Some become social misfits and need constant support, although those who continue treatment have an excellent chance of success.

Long term psychotherapy, combined with the right drug, produces the best results. It strengthens weaknesses and gives a sense of security and self-assurance that fights the disease. Drug therapy maintains the equilibrium of the patient at a level receptive to psychotherapy. Without its calming effect, the disturbed mind rejects communication. However, psychotherapy is ineffective in some cases and has been known to aggravate symptoms, and cause the disease to accelerate into total disintegration if pressed improperly.

It is clear that we need better techniques with which to treat schizophrenia. Prevention is the real answer. One approach is to change the body's chemistry so it does not produce excessive dopamine and/or too many dopamine receptors. This would prevent the disease instead of having to treat an unbalanced brain. Genetic counseling of couples contemplating marriage is one answer. Gene replacement or brain cell transplantation is another, neither of which is now available.

COPING WITH SCHIZOPHRENIA

When you have schizophrenia, how do you cope with it? Your problem is that you don't see things as others do. What you see as real is not what they see. You are at your best dealing with material things, such as books, written words and mechanical articles, so it is wise to choose work that interests you and that you do alone and in which you may have a talent. So choose the work you like and that will help you develop a sense of self worth, for you have a condition that will be with you all your life. There are many rewarding jobs that don't involve people. There is writing, music, painting and other forms of art. There is bookkeeping, accounting, research and repair work. As you are a thinking person, you are apt to be very good at these vocations.

When you are with people your reasoning powers are at their worst. This makes you tense, resulting in even greater inability to understand facts, so as you have trouble getting along with people, pick friends you feel comfortable with. All this will come through psychotherapy and as you mature.

You are sensitive to criticism, unduly critical of yourself and filled with guilt at not being adequate, so it doesn't take much to upset you. First, you are remorseful and then angry, and if you have a paranoid tendency, it can get out of control and make you aggressive. But don't feel guilty for being the way you are. You didn't bring it on yourself.

If it is suggested that you do this or that, give it a whirl and find out if it works. However, don't accept every suggestion, examine each one carefully. And remember that your interpretation may differ, but don't reject it, discuss it with those who suggested it. You know that they are more apt to have logical thinking than you do.

Trust those who love you and listen to them. And remember, it is essential that you take the medication prescribed by your doctor. It is also important to stay away from drugs, such as alcohol, nicotine and caffeine, which you might be tempted to try. Exercise instead, it always makes you feel better. And remember that there are many things that can help you lead a comfortable and full life.

HOW DO YOU COPE AS A FAMILY?

When one of the family has schizophrenia, you also are genetically the more thinking type, therefore sensitively attuned to the patient's symptoms. Although this illness is sometimes cured, the majority of cases are not, and you have to accept that your child, sibling or relative will have symptoms through life.

Let us assume that the patient is your son. He was normal until about eighteen, when he began to talk strangely. One day he told you the telephone had disappeared although it was right there. Another time he announced that his heart had settled in his stomach. He went to buy a newspaper and said that the neighbors were following him and laughing. He was so tense and apprehensive, you took him to the doctor. The diagnosis was schizophrenia and he was placed on medication. This relieved his delusions.

There are times when you are sympathetic and understanding, and times when you are angry, because he won't listen to you. Then you may reject him, but this is when he most needs support. Remember that he hates himself and his symptoms and is hypersensitive to criticism. The slightest sign of rejection can send him into a tailspin.

What sensations does he experience? His hallucinations are usually auditory, but may be visual and vary from unpleasant to ridiculous. One patient, who adores his wife, hears a voice telling him to kill her. So what your son hears is irrational and unrealistic, and this confuses him. He has trouble reasoning, and is apt to be naive, so he finds it difficult not to believe the voices he hears. Sometimes he thinks it is God speaking, or the President of the United States or another important person. His other senses may not function as they should, such as smell and taste. This is very real to him, and his weak self-esteem and poor reasoning make it impossible not to trust his senses. He may also have delusions, such as thinking he is Jesus Christ or a famous character. These delusions are fixed and he cannot see otherwise, despite sensible arguments to the contrary.

So what more can you do for your son? Persuade him that it is essential that he take his medication and go regularly to therapy. This alleviates his symptoms and improves his understanding of reality. Do the best you can to steer him away from drugs such as alcohol, nicotine, caffeine and of course, hard drugs, especially psychedelics like pot and LSD. Encourage him to exercise. This is very important. Be convinced that he is sure of your and his family's love, and show this in every way possible and say it repeatedly. Bring kindness and devotion into the picture as well, but do none of these things to an extreme that spoils your own life, for you are a person too.

Psychotherapy will help you understand his condition so that you treat him as you should, and it is important that other members of the family undergo therapy as well. The balance to love and understanding is firmness and discipline, so he knows that you won't allow him to use or manipulate you. A tight rein and strong hand is his security and helps maintain his stability. If the scale is out of balance, he is miserable, and his illness worsens.

Guide him firmly and gently into an occupation that involves things, not people. He can operate well alone if allowed to go his own way in his own fashion, and he can enjoy his work if he is not subjected to demands and criticism. You may be frightened when he says he is receiving messages from space, or can detect vibrations from people and interpret them, but he can carry on with his work smoothly as long as he is allowed to function in his way.

Create the environment your son lives in. This contributes to his comfort and yours, and gives you the satisfaction of doing something to help. But you must realize that this has no effect on the disease, no matter what is done.

Sometimes patients get well spontaneously for reasons we do not know. But those who don't have a tremendous impact on their families and everyone in contact with them. They misinterpret and distort what they see and hear and react strangely to situations that are perfectly normal. You treat a patient kindly, give him a compliment, and he says: "I can tell you hate me." This throws you off and you wonder what you said to hurt him. If you say you are sorry, you play into his hand and fit into his distorted world.

You must expect this kind of reaction, and it is important to reply with a clear and consistent attitude that is intellectually and emotionally correct so the schizophrenic can observe what realistic thinking is. Don't go along with his bizarre way of thinking, but get him to go along with yours. He is trying to build a fantasy world and place you in it. Don't let him.

Families learn through therapy to understand this illness and not be critical or hostile, but supportive of the schizophrenic as a person with a disability and not someone to be belittled or made fun of. Here is what not to do. A schizophrenic boy got into an argument with his mother and sister over cooking. He claimed cooking is not difficult, even he can do it. He was trying to build himself up. They attacked him and said he could never learn to cook, and who is he to talk that way when he gets free meals and doesn't have to lift a finger? This upset him so he admitted he was no good and decided to kill himself. The boy was trying to find a niche in the family and when he was rejected, he acted out.

Schizophrenic families are apt to be very schizoid, which is the medical term for extreme thinkers, and the type of altercation described above takes place in all of them. The paranoid tendency surfaces and dominates the relationship and the weaker member takes the brunt. Psychotherapy helps families learn how to support the ill person and weather these storms. This boy should be told that his ideas are interesting and he has a right to them, although his mother and sister don't agree. But families feel threatened by criticism as does the schizophrenic and often condemn opposing ideas because of their own rigidities and prejudices.

Schizophrenics differ widely in what they experience and how they act. One extreme is the intensely disturbed person, wildly excited, hallucinating, laughing inappropriately, talking bizarrely. The opposite is a normal slightly depressed or slightly euphoric person, with thoughts that are never revealed, but which are as bizarre, inappropriate, unrealistic and impossible as those of the person who displays his. Outwardly, he is an ordinary person, but inwardly he is in turmoil.

Here is an example. A student was the star end of the football team at the University of Nebraska and was expected to be the top player. While on his family's farm during the summer he took a rifle, went back of the barn and killed himself. He had just had breakfast with his father and mother and appeared perfectly well. He was an A student, a high achiever, well liked and an all around boy. Why did he kill himself? He is the type of schizophrenic who has terribly disturbing thoughts, keeps them bottled up and suffers in silence. He can't let them out and, being extremely sensitive and self-critical, makes a false decision which he fusses and frets about until it is blown up out of all proportion. He sees no solution except to end it all.

As you can see, the symptoms of schizophrenia are widely spread, and so is the variety of manifestations in the different types. There is the excited catatonic schizophrenic and the passive catatonic, the residual and the paranoid. The residual has no delusions and does not hallucinate. He is just dull, flat, empty. His thinking is not bizarre. Some doctors claim that most prostitutes are residual schizophrenics, girls who have little feeling. They are just empty. They don't care. To them, prostitution is the easiest way to make money because it doesn't require thinking. The emotion of sexual relations is lacking, so they can function without a need for love or a sense of guilt. But they try to obtain some feeling through sexual acts.

The paranoid schizophrenic is suspicious, fearful and constantly misinterprets the words and actions of other people. He can also be hostile and critical. The catatonic schizophrenic is slow, irresponsible, negligent and lags behind, giving a million excuses why he does nothing.

Some cases don't fit any category. Here is an example. A young man in his early twenties had always had top grades and went on to graduate school where he began to slow down and drift out of his usual pattern of behavior. He was due to receive his masters degree five years ago but is still studying. He cannot seem to get hold of life and is unable to function except to study. This is typical of schizophrenics. They can devise brilliant concepts, have many capacities and potentials, but cannot produce. Their "doing" capacity is nil. What it amounts to is all talk and no action. It is tragic.

As we have explained, there are five categories of this complex disease. (1) disorganized, (2) catatonic, (3) paranoid, (4) undifferentiated, (5) residual. The above case does not fit any, so is loosely described as "other types, the waste basket". He is not disorganized because he is very organized in his studies, he is not paranoid, he has the average amount, he does not fit into the undifferentiated group because his case is clear cut. He can think but cannot act. His is not a residual case because he has always been this way and his illness is progressing and interfering more and more with his life. He is not catatonic because he is vigorous and not lethargic in any way.

This patient has high energy and high intelligence, and because of this is able to overcome a lot of the basic effects of the disease which a person with lower energy and intellect could not.

There is new hope. Although there is no known cure, five recent research studies show that up to fifty percent of schizophrenics get better spontaneously and are able to live adequately without treatment. We have also learned that drugs alone are not as effective as with psychotherapy, and that intelligent management of the daily life of the schizophrenic is equally important.

We have known since the discovery of the neuroleptic drugs that many of those afflicted with schizophrenia can lead relatively normal lives, if they continue to take the medication prescribed by their doctors. But there is a strange quirk in the schizoid mind, the more thinking mind. They militate against taking drugs. They don't even want to take aspirin or Tylenol, much less drugs for their delusions and hallucinations. The women complain of side effects such as weight gain, which is true, because the drugs slow the metabolism, and the men complain of poor sexual function. They all complain of lethargy and lack of interest, even though their minds are much clearer when on the drug.

So they all object to the drug, but never about hearing voices or the other disturbing symptoms. When asked if they would like to have the voices stop, they reply by beating around the bush and saying: "When I hear them, I am happier than I am now. I can talk to the voices, they say things I like to hear and I am in my own little world."

We never know what a patient might hear. When they have what is called a "directive delusion or hallucination", that could be a dangerous symptom as in the case of this forty year old man. Stephen ran a successful business, but socially he had difficulty establishing relationships with women. When he sought medical care, he stated that whenever he dated a woman he heard this little voice say "Kill her, kill her". This made him nervous as he claimed he would never hurt anyone, and decided that perhaps the voice was telling him that he is not really interested in that girl and should not take her out again. Stephen never married, although he dated a woman for five years or so and established a healthy relationship with her.

This is an example of a directive delusion and we never know what it might lead to, but seldom has a schizophrenic hurt anyone. One doctor had two cases with that kind of admonition among hundreds of cases he had treated, but he never reported any action taking place despite the voice's order.

The mental, emotional and financial toll of this condition is enormous and reaches out from the core of the family, where the worst damage is done, to the community and society as a while. Schizophrenics occupy half the beds for the mentally ill and retarded and one-fourth of-all hospital beds in the country.

Sixty-five percent of schizophrenics live with their families after hospitalization and the strain is great. Parents and siblings pay a terrible price from their demands and are despairing and exhausted by the endless obligation. The patient creates constant turbulence and the family is lost in an engulfing, thankless process that is often hostile. Health professionals consider families the "forgotten group".

A million families struggle with this burden and are called upon to be doctor, nurse and social worker, while the victim loses reality in delusions, incoherent thoughts and bizarre behavior that are as frightening to him as to those around him. Then he may withdraw into apathy (catatonic stupor), and be unreachable for days, unable to care for himself.

Parents watch their child tormented by hallucinations and, unable to shake them off, lose control and deliver an onslaught of uncooperative behavior. Love and patience is sorely tested many times. A hostile retort or attitude on the part of the family increases the illness, creating guilt and anger.

Nevertheless, there is hope for schizophrenics and their families. Modern biotechnology is galloping ahead, with the support of computer technology as an aid to research on the brain. Heredity appears to be a factor in the etiology of the disease and the tools of molecular biology will make it possible to identify the gene responsible.

CAT scans (computed axial tomography) of the brains of schizophrenics show that the prefrontal cortex is underdeveloped. Comparisons are made with normal subjects, and pictures, taken during tests that require concentration, show that their brains don't have the same activity as normal brains. There are also abnormalities in the structure, most notably abnormally large ventricles, which are the hollow spaces in the brain filled with fluid. Thus, schizophrenics have less brain tissue. Studies show that this is related to the disease and does not happen over time, and whatever causes these enlarged ventricles does so before symptoms appear, and does not change with or without treatment.

A NEW MANAGEMENT OF PATIENTS

Results from a thirty-year study of the management of the schizophrenic culminated recently in a three pronged approach to the care of these patients. It is called "family management" and is cropping up all over the country and proving to be very effective. Patients are treated with drugs and psychotherapy in the hospital and then released to this program, which closely monitors their lives. The primary purpose is for family and friends to create an atmosphere of calm and support, the concept being that a psychotic person living with normal people becomes healthier. Thus, living in the proper environment can offer therapy not available from psychotherapy or drugs, and results in a reduction of high doses, reducing side effects. Two-thirds of patients in this program live normal lives, and one-half no longer have symptoms.

There are three types of management programs. One, family management, whereby family members are instructed how to create an atmosphere in the home, how to monitor the patient, and cope with day by day difficulties. Each program is tailored to the individual case and all work well.

The demands of a patient can conflict with family life and create restrictions difficult to tolerate. An example is the insistence by one patient to cover the television screen "to keep the eavesdroppers out." The family should have relief from the presence of the patient, especially when entertaining friends. He can stay in his room when company comes. So he must agree to accept discipline, and discipline is a form of security, which the patient needs. The family is also entitled to their lives outside the home. All of this contributes to maintaining an environment conducive to helping the patient recover.

The second type of management program is community services for patients whose families don't take them in. These consist of enrolling members of the community, on a volunteer or paid basis. They can be neighbors, friends, kind hearted individuals and relatives who form a crisis team for one patient. Someone is on call twenty-four hours a day. In communities where this program is ongoing, there has not been one case of serious relapse.

Three, one community created what is calls "network therapy". This begins with a meeting of people the patient knows who are willing to help. The patient's story is told and what is needed to support him. Each member offers help in some way, such as driving or shopping, helping with financial affairs or being on call for a crisis.

This type of management has proved to be successful beyond expectations. Research shows a one year relapse rate, meaning hospitalization, of under ten percent. Patients receiving drug therapy only, had a rate of fifty percent, those with no treatment, sixty to eighty percent during the first year after discharge from the hospital.

Often circumstances make it impossible to take the schizophrenic into the home, and some families are not willing or able to do so, and there are patients who have no family for whom there are programs that emulate family management. However, few can afford them.

The discovery of the miracle drugs to treat schizophrenia in the 1950s. was a mixed blessing. Patients were treated in the hospital and discharged to their families, or into society to fend for themselves in the belief that they could. Many patients have no place to go after they are discharged. If their families don't want them, society doesn't either, nor are all communities motivated to set up management programs. Consequently, many schizophrenics are street people.

Here are two cases.

A woman saw a homeless person on television and realized to her horror it was her brother. She tracked him down and asked him to come home. He refused, saying he had friends on the street and didn't want to leave them. She asked if he was taking his medication and he said, "No, I don't need it".

A psychiatrist met a former patient on the street. He had not seen her for ten years. She was hearing voices and talking back to them. The doctor asked if she was getting treatment. She replied that she was fine and didn't need it. "The voices are wonderful, I love them", she said, as she walked off babbling to herself.

Schizophrenics have poor judgment and are not able to plan or predict their needs today or for the future. About twenty-five percent are so ill they are incapable of functioning, even in the setting of a caring family. They require permanent placement in an institution such as existed before the discovery of the neuroleptic drugs. Schizophrenia is the cancer of mental illnesses and the object of stigma and ignorance as was cancer fifty years ago. The government spends $300 on research for every cancer patient and $17 for each schizophrenic. Two million Americans are afflicted and the cost of treatment and lost productivity is estimated at forty-eight billion a year. In 1987, the Congress increased research funds for schizophrenia by five million. More is needed. Also needed are scientists who choose research in this area.

At this time the causes of schizophrenia are unknown, but there are several theories being studied. At a recent meeting in West Berlin, one scientist summarized new findings as three sure facts: (1) it runs in families, (2) certain drugs that influence the brain's dopamine system make it better and (3) there may be structural abnormalities in the brains of schizophrenics.

(1) The tendency to schizophrenia is inherited, and there is evidence that it is similar to coronary heart disease in that it tends to run in families. However, the genetics are not clear cut like Huntington's chorea, for which a single gene has been identified.

(2) The dopamine system in the brain is receiving a good deal of attention. Of the approximately fifty drugs that relieve the symptoms, all work through the dopamine neurotransmission system, however, if that is the cause of schizophrenia, drugs will not necessarily correct a defect in the brain. So far, research points to there being an increased number of receptors for dopamine, rather than higher levels.

(3) Schizophrenic brains examined at postmortem reveal changes in regions of the brain called the hippocampus and amygdala. So there is a possibility that the disease is caused by an abnormality in the structure of the brain which could come from a retrovirus or a congenital defect from a defective gene. Another theory suggests that a retrovirus may have entered the human genome generations back and been carried on to subsequent generations. There is no hard evidence, however, to support this theory.

It has been observed that one-half of schizophrenics are born in the winter or spring when viral infections are prevalent. A recent study sheds light on the speculation that a virus may be involved in some manner.

In 1957, there was an epidemic of influenza in England and a review of birth records showed an abnormally high rate of schizophrenia in people born four months after the peak of the epidemic. Of the persons born between February 25th and March 15th 1958, forty eight were hospitalized for schizophrenia between 1976 and 1986, compared to twenty-five born during the same month in the preceding and following two years. This differential of eighty-eight percent would arise by chance only once in ten thousand times. Results of research in Finland confirm the same findings.

Although it is not possible to determine whether these mothers of schizophrenics were infected with the influenza virus, it is known that women who contracted flu during that 1957 epidemic were twice as likely to have babies with congenital malformations. It is suggested, therefore, that exposure to influenza during pregnancy may affect the development of the child's cerebral cortex.

The study began with one hundred and thirty-one patients who were first examined in 1968 and who had been ill for about two years. They were interviewed after two, then five and finally eleven years, when only forty patients could be found. They were rated on the basis of how long they had spent in mental hospitals, how often they had social contacts, the ability to work and the severity of their symptoms.

The final results of the eleven year study on this group of schizophrenics reveals that the severity of the disease changed little after the first few years. The study also provides information as to what symptoms and characteristics may predict a good prognosis. Patients considered to have a good prognosis at the first evaluation were doing better after eleven years on all points. Patients who had spent the least time in hospitals and had the most social contacts were also, as were those with symptoms, such as apathy or emotional numbness, usually considered bad prognosis signs. However, they were not associated with a poor outcome at the end of the study.

Those patients who had delusions, disordered thinking, or hallucinations and showed less severe symptoms at the follow-ups, were not better off. Patients who were anxious and depressed at the beginning interview did not end up especially well after eleven years, but patients who scored high on a more general measure of distress ended up with a good outcome.

These findings are contradictory, and the authors suggest that acute psychosis distorts the manifestation of moods and emotions, so that the observing scientists become confused and unable to interpret the symptoms accurately. Of enormous help in studying the biology of the brain are the new imaging techniques, the MRI (magnetic resonance imaging), PET scans (positron emission tomography), CT scans (computerized tomography) and BEAM (Brain electrical activity mapping), an advanced version of electroencephalography (EEC). A computer analyses the activity recorded by electrodes placed on several parts of the skull and presents the result on a screen and show brain wave responses to stimuli. Measurements of blood flow and metabolism, receptor mapping in living patients and the use of molecular techniques to discover genetic effects are also extremely beneficial.

Debate continues as to whether schizophrenia is one disorder or a component of a cluster. This cannot be determined until molecular genetics studies are done. Other researchers hold that all mental illnesses are related and form a spectrum, with depression at one end and schizophrenia at the other.

One scientist summarizes this data: "A complex interplay of genetic and environmental factors determines the development of schizophrenia." All scientists in the field have arrived at a consensus that schizophrenia is of overwhelming complexity.

7 DELUSIONAL PARANOID PSYCHOSIS

The words paranoia and paranoid appear frequently in conversation. As already discussed, paranoia is an important personality trait and can be exaggerated in mental illness. There is paranoid schizophrenia, depression with paranoid trends, paranoid states, mania with paranoid trends, paranoid reactions, delusional paranoia and so forth. When the word is used to define a disorder, such as paranoid schizophrenia, it means that paranoia is a strong element in the illness.

Delusional paranoid psychosis is as old as the hills and was recognized in man and described before Hippocrates. About the second century A.D., it disappeared from medical literature, and reappeared in the eighteenth century, defined as "an insidiously developing, unchanging, delusional system, accompanied by clear and orderly thinking without hallucinations." This remains the definition today. It is a rare disorder that develops slowly and becomes chronic and is extremely complicated with an internally logical system of persecutory or grandiose delusions. It does not interfere with the remainder of the personality which continues intact.

The personality of a delusional paranoid psychotic is involved and isolated, and may be related to schizophrenia. It comes out in mania and may be linked to the doing personality, the psychopath, or it may stand alone. Some textbooks don't call it a psychosis, but a way of thinking.

Delusional paranoid disorder is an extremely uncommon psychosis. It shows severe brain dysfunction and is a chemical disturbance of thinking. Both doing and thinking personalities can be afflicted. Patients display greatly exaggerated signs of paranoia, extreme fearfulness and suspicion, unreasonable hostility and illogical thinking. They cannot take criticism, they place all blame elsewhere and refuse to assume responsibility for anything that goes wrong. There is no deterioration of personality as in paranoid schizophrenia.

Studies reveal that delusional paranoid psychotics have certain personality characteristics in common that are well defined. They are hardworking, self-driven and conscientious, compulsive and fanatical, especially in the areas of politics and religion. They are insecure, and fear that someone, anyone, even those closest, can turn against them. They seldom trust anyone and live in a state of anxiety. They keep to themselves and make few friends. Delusional psychotics seldom marry, and when they do the marriage is rarely successful. They are always watching what others do and interpret all actions as directed toward them.

They don't consider their thinking distorted, nor do they understand why others don't think as they do. They may be socially acceptable, but a mask of arrogant self-confidence hides a weakness. They go to great lengths to defend their positions, and never blame themselves, but are highly critical without cause and exhibit unwarranted and irrational hostility. Thinking is cluttered with denials and assumptions, evading responsibilities and passing them on to others. This mechanism of evasion and placing blame is called "projection". Symptoms are mildest and most constructive when the patient is young. Older patients become steadily more hostile and easily provoked.

This sounds like the excessive paranoid personality described before.

How are they different? The line is drawn when a person's behavior is so severe that he is unacceptable to family and associates, or is so disturbed by delusions of persecution that professional help is mandatory. There are five types of this disorder and some cases may present more than one. The erotic type focuses on the allusion that he or she is loved by another person who is usually of a superior status or a famous person who is looked up to, even a stranger. The theme of the infatuation is romantic love and ideal spiritual union rather than sexual attraction. The deluded makes efforts to get in touch with the object of adulation through telephone calls, letters, offerings of gifts and visits. Occasionally, none of this is done and the person worships in secret.

Erotic delusions are fairly common and a source of harassment to prominent figures. One patient believes that a celebrity is in love with her and awaits a sexual proposition. She fantasized that she will be indignant, while secretly thrilled and filled with pleasurable anticipation. This form of delusional disorder relates to erotic thoughts with no actions involved. It is usually associated with women and passive men.

The *grandiose type* is fairly common. This person can believe sincerely that he or she is serving the President of the United States. These people usually experience delusions related to their line of expertise, but the most severe cases choose an area with no possible link, or only a fantasy tie. They often are convinced that they have a great talent or important discovery yet unrecognized, and want to take it to the Federal Bureau of Investigation, the State Department or other important government agency. Less common is the delusion that the person has a close relationship with a prominent person or movie star or other celebrity.

Sometimes the delusion may be a religious one and the individual could become the leader of a cult.

Delusional paranoid psychotics may have a pure form and experience delusions of grandeur associated with their capabilities. They are highly intelligent, very strong and free of misconceptions when at work. They have been known to contribute with genius to high level research and development that is later carried out by peers. This kind of psychosis is extremely rare and is considered incurable.

The *jealous type*. In this case, a person is convinced for no reason whatsoever, that his or her spouse or lover is unfaithful and having an affair with another. To substantiate this accusation, he or she goes about collecting evidence, such as spots on the sheet, misplaced clothing, and anything that can relate to a possible infidelity. Then he or she confronts the spouse or lover with accusations and takes steps to keep the accused at home or from going anywhere unaccompanied. He or she may follow the person suspected to be involved, launch an investigation and occasionally physically harm that person. Here is a case in point.

John, thirty-nine, a lawyer, was convinced his wife was having an affair with her obstetrician. She kept praising the care he gave her while carrying her second child, and her husband suspected there was more than professionalism between them. When the baby was born, John received a modest bill and this confirmed his suspicions that the child was not his.

This patient suffered from severe insecurity and had misgivings about his potency. During his wife's pregnancy he had an extra-marital affair, for which he felt guilty and afraid of being found out. He projected this onto his wife, reinforcing the certainty of her unfaithfulness. This delusion persisted for months despite evidence to the contrary, when his intelligence led him to professional advice. Psychotherapy helped him work out his paranoid tendencies.

The *persecutory type*. This is the most common type of delusional paranoid disorder. Its range is wide, from the simple theme of being cheated out of a few dollars to an array of accusations of being persecuted. These people are convinced they are conspired against, spied upon, followed, poisoned or drugged, maligned, harassed or prevented from pursuing their goals. The slightest rejection becomes blown up to vast proportions and often leads to legal action against the adversary. The frustration of not obtaining what he or she considers justice creates anger and resentment and may lead to violence. Here is an example.

Adam, forty-nine, is brought to the doctor by relatives after admitting that he was going to kill himself to escape being tortured to death by gangsters. He is obviously badly frightened, but polite and cooperative, and tells his story intelligently and willingly, not aware that he is suffering from delusions. He has never worked, and lives in a hotel on a modest income from investments. He occupies himself talking with a few friends and reading the newspapers. He likes to bet on the horses, and places daily bets through bookies by telephone.

His acute paranoid attack was triggered when a horse he had bet on came in first and he went to pick up his money. He was told he had made a mistake, his horse had lost. He was upset, went to a bar, thought it over, and decided that he had been cheated. He returned to the bookies, furious, and demanded to be paid. When they refused, he insulted them and invited them outside. They refused again, and he went back to the hotel raging mad and sat in the lobby, brooding. He came to the conclusion that the bookies are gangsters and he had done a dangerous thing in challenging them.

This decision frightened him and he was sure the gangsters would pick him up, torture him and kill him. So he fled to a cousin's house a thousand miles away, all the time imagining that strangers along the way were the gangsters. His trip took several weeks as he hid in terror much of the time. He was convinced he could not escape his pursuers and decided to kill himself. On arrival, he wrote a suicide note and waited for a moment alone. His cousin discovered the note and took him to a psychiatrist.

The *somatic type*. There are several forms of this kind of delusion. The most common is that the person has a foul odor coming from his or her mouth, skin, rectum or vagina, or that there are insects crawling over the body or invading the intestines. They may consider that their bodies are ugly and misshapen and that everybody sees this and says nothing out of politeness. These delusions are tragic, as they relate to an individual who is perfectly normal.

Delusional paranoia can develop in a family or group living together. Two or three members develop symptoms simultaneously, as though they had caught it from one another. This occurs most often in religious cults that encourage blind allegiance. Recently, a case was in the newspaper of a men who jumped off a high building to his death, claiming that he was joining God. His wife threw their five children after him and then herself.

What are the causes of delusional paranoid illness? The paranoid personality is particularly susceptible. Observation and experience have proved that it is hereditary, and there is high incidence in persons who have difficulty in establishing good interpersonal relationships in childhood or adolescence.

There are several causes being considered by scientists. Excessive dopamine and/or dopamine receptors, as in schizophrenia, an excess of norephinephrine, as in manic depressive psychosis, thyroid and parathyroid excess, abnormalities of the pituitary and other endocrine glands. Arteriosclerosis of the brain, kidney disease and conditions that reduce oxygen and nutrition to the brain are suspected. The abuse of drugs, especially stimulants, can also cause this illness.

What is the treatment? A correct diagnosis must be made first and this requires a thorough physical as well as a psychiatric examination. If a physical condition is present, steps are taken to reverse it. The psychosis is treated with anti-psychotic drugs, the phenothiazines and others described earlier. Once the brain chemistry is stabilized, psychotherapy is begun, and is of great benefit to the patient in understanding his illness. Except in pure paranoid psychosis, the prognosis is good.

What is the incidence? Women and homosexual males are the most susceptible. Living in a heterosexually dominant world, homosexuals show more paranoid traits and have more paranoid trait genes from the male parent. The onset is usually during middle life. The incidence is low in the population, but the influence of one paranoid can be so pervasive that the low percentage is magnified by his extreme drive to control.

The course of this illness varies widely. In some cases, symptoms disappear within a few months without ever appearing again. Others become chronic, especially the persecutory type, but in all cases the waxing and waning of symptoms are common, with periods of remission followed by relapses.

As far as daily functioning is concerned, the majority of cases lead relatively stable lives between the episodes of illness. However, while their hours at work may be satisfactory, their personal lives are not. Living with a delusional psychotic is extremely difficult.

How do you cope with the person with delusional paranoid psychosis? Here are some suggestions. He is the most difficult person to live with. He is enormously sensitive and very bright. He remembers everything and his memories are acute, particularly for unpleasant events. If you question him, challenge him or tell him he is wrong, he will never let you forget it. If you insult him or stand in his way, or even inadvertently cause him distress, he will throw it back at you over and over again. This psychotic is a victim of his illness, which is a thinking disorder that becomes more and more intense.

Most people remember pleasant happenings, but the delusional paranoid, for reasons unknown, recalls disagreeable things more acutely than everyone else. So if you have to live with one, there will be many unpleasant incidents. One day, Edward was cutting his fingernails and said to his wife, "You say you love me, but why did you slam the car door on my hand fifteen years ago?" An ordinary person would have forgotten this but not Edward. And when you explain that you didn't do it on purpose, it was an accident, he replies, "That's a weak excuse. You were really out to hurt me, because you didn't want me to play golf and leave you at home!" To him this was a personal injury and he builds up a plot and can't let it go.

So the delusional paranoid psychotic has all the characteristics of the paranoid personality greatly exaggerated. It is a kind of malignant intensity and reality is distorted. He is convinced that you are out to get him and every little thing you do is interpreted as potential harm to him. Such as, "You've always hurt me and I can tell by the look on your face that you hate me and are planning to kill me!" Or "I can tells because you don't look at me much any more, that you're plotting something against me, and I don't know what I am going to do!"

What happens to you and the family? You all become worn down and terribly frustrated, because you are constantly inhibited trying to deal rationally with irrational concepts. Anything you say or do is misinterpreted as an attack. If you smile or don't smile, laugh or don't laugh, frown or cry or don't, it is wrong in his mind. So he is generally disliked and has no friends.

So how do you manage living with this person? Keep a consistent, firm position, logical and orderly, and unable to be swayed by the distortions you are faced with daily. And when he says, "You are looking at me in a funny way, and I don't like it!" Answer, I'm not looking at you in a funny way. That's your problem. I am looking at you as I always do and caring for you the way I always have and always will."

Be steadfast and consistent in a loving, caring, supportive attitude, no matter what is said to you. This takes strength and wisdom and you have to realize that this person's thinking is distorted and you must not allow yours to become distorted too. This constant awareness is wearing and you need a break from time to time. Be sure to do this frequently, because you are under constant, tremendous pressure.

You have another responsibility, and that is to see that the patient continues treatment, taking the medication and keeping the doctor's appointments. This is difficult because he doesn't consider himself ill. He thinks everyone else is.

The delusional psychotic is also apt to be hyperchondriacal. He exaggerates every little complaint, the slightest headache or backache or touch of indigestion. The smallest physical distress is magnified and becomes a terrible burden to him and he worries until he is really frightened. Then he needs a lot of support and reassurance.

There is only one person in his world that the delusional paranoid trusts. It is rarely a group, like the doctors at the clinic or his family, it is usually one person. Anyone who tries to control him is like a red flag to a bull. And it is the same if he feels threatened. The way to handle him is to go along with his ideas and agree with him. You need not act upon them but let him think that you will. He is usually not harmful, is always concrete, but inappropriately so. If you try to block him or threaten him, he becomes abusive and potentially harmful. When a psychiatrist does not understand this type of patient and fears him, he is subject to assault or even death. Doctors have been shot or knifed by patients when they handle them the wrong way. Paranoid personalities, from the mildest to the psychotic, will stop at nothing. They are all capable of attacking and causing injury, sometimes only verbal or written, such as a "poison pen" letter, up to murder. When they feel threatened, they are willing to take any action to protect themselves, and the more psychotic they are the less reality they deal with and the more dangerous they can be. They are strong, forceful people, vigorous and convincing, often hypomanic, sometimes manic, and there is a link between mania and delusional paranoid psychosis. However, their assumptions are false.

It takes a certain personality to treat paranoid psychotics and I, the author, have done this for over twenty-five years and never been injured. This is because I am a non-controlling doctor and go along with their line of thinking. They can see that I am not trying to make them change their minds. I have had them bring weapons into my office, pull bayonets out of their boots and give them to me. I have had them reveal plots that are life threatening, but I have never been attacked. They are almost like pets, and seem to have an insight of who is going to pet them and who is going to kick them, an almost primitive sense of danger. They have an inner alarm system that goes off in tense situations, and if you create tension, it goes off, and they may attack because they think you are going to attack them.

I have a patient who cannot understand why no one likes him. He says I am the only person who understands him and why can't others? He calls me up and says, "You know I am right." I don't know what he is talking about, but I will support him because he has to have a friend on whom he can depend and who will accept him and his ideas and opinions. Then he can say, "He is my friend". Unfortunately, he has no other. Then, on the basis of my encouragement, he can work safely with me to resolve his fears and self doubts.

One of my patients is an eighty-five year old woman who tells me: "You are my only hope. You are the only one who can go to court and get this conservatorship off my back, because I want to run my own life and do the things I like. The court has taken my money away from me, and I want to control it. My parents gave it to me so why can't I have it?" I go to court and testify that she thinks logically but with a false premise. This causes the court to consider that she can manage her money wisely. But she can't. She already lost a good deal through unscrupulous people and so-called friends who claim they will do her a service if she will give them money. They promise to buy pills for her she should not have and arrange trips she should not take. If she regains control it will soon become apparent that she cannot cope and her conservatorship will be renewed.

Delusional psychotics don't know their own best interests, but are convinced that they do. They must control but will not be controlled. Their method is blatant and obvious, using threats, sometimes sexual, or physical force, but most often by manipulation through financial, political or religious influence. They are logical thinkers, but their logic is based on false assumptions. They build a huge body of truth, supported on erroneous conclusions. But you can't find fault with the logic of their thinking or the common results of their aggressive style, such as the gain of power, money or control.

8 MANIC DEPRESSIVE PSYCHOSIS

Like schizophrenia, manic depressive psychosis is not new. It was described in the earliest Indo-Germanic cultures and the Homeric epics. Hippocrates wrote about mania and melancholia, which he considered chronic conditions, but he made no connection between them. Several centuries later, Aretaeus, a Roman physician, described a connection that linked mood extremes. He wrote so accurately that it compares to today's views. He also noted that although the two are related, mania doesn't always precede or follow melancholia, and he outlined in detail the depressed attitudes of the melancholic and the hyperactive behavior of the manic.

These ancient teachings were either forgotten or ignored until 1851, when the condition was again described by a French psychiatrist. Since then it has been recognized as a mental illness.

Manic depressive psychosis is a disturbance of mood and, like schizophrenia, is caused by a defect in protein metabolism. It may also follow a metabolic disturbance of the thyroid or other hormones that affect the brain, but neurotransmitter defect (metabolism) is the usual cause. It originates in a recessive gene, and is characterized by alternating attacks of elation (mania), and depression (melancholia), ranging from mild to severe. It is called bipolar because the mood reaches both poles, high and low.

What are the symptoms? There are recurring attacks of depression and elation, or mania, and if the manic episode lasts a week, the depressive phase may also. Symptoms show up in many patterns, and last for a few days to up to a year or more, following closely upon one another in some cases, or spaced between periods of normal mood. A few patients never have a normal mood, but are always either high or low.

The diagnosis is not difficult, however, a medical and a psychiatric work-up is advisable. As with schizophrenia, chemical and psychological tests are of help, and recently chemical tests of the brain were developed to diagnose all mood disorders. It is easier to make a diagnosis during the manic phase, for the depressed state may be confused with reactive depression or a physical illness. A devastating life event or physical injury can trigger the first attack. It isn't caused by this, however, for the patient was already in a prepsychotic condition which would have been set off sooner or later.

The depressed phase may begin suddenly, or creep up slowly without the patient being aware. He has been depressed before and assumes it will pass. But the symptoms increase in intensity, and he cannot think clearly or work effectively and his appetite and sleep are disturbed. Often it takes a major effort to get out of bed in the morning.

Everyone experiences depression, due to an unhappy life event or the usual mood swings from changes in brain chemistry. These are normal and based on environmental or causes from within, called endogenous. But the feeling of depression associated with this psychosis is different. It comes as a payback of energy reserve which has been drained by the manic phase.

Patients have physical complaints as well. They may describe a heavy feeling in the chest, numbness of body, and an inability to eat, sleep or cry. They are unable to move, and have vague aches and pains and stomach problems. They feel unworthy and seventy-five percent think of suicide. However, only ten to fifteen percent try it and most are thwarted. Young mothers in psychotic depression think of their children as extensions of themselves and plan to take them along in death.

A depressed person has morbid thoughts. His phobias are accentuated and obsessive worries flood his mind. He thinks slowly and can't concentrate. He has to push himself to perform the simplest tasks, neglects his grooming, and wears the same clothing day after day. He hates himself for his condition, the neglect of his work, and letting his family down. He blames himself, not the illness. It is obvious that he is slowing down. He walks at a snail's pace, is physically weak and sexual desire is severely diminished. He is indifferent to others, refuses to leave the house or talk on the telephone, even to family and friends.

There are different degrees of mania during the manic phase. The mildest is hypomania, then acute, then delirious. In hypomania, the person appears to be normal, although high in mood and excitable, until he does something completely out of keeping. He is a great accomplisher, up to a point. In business, he promotes innovations that advance him and his firm. The artist feels creative and produces more than usual. The blue collar worker pushes himself more than his fellow workers and is more dedicated. However, hypomanics are characteristically hostile, suspicious, self-protective, demanding and controlling, all signs of paranoia that is accentuated during an excessively high mood.

In acute mania the symptoms are more pronounced. The patient becomes restless and irritable. He is unruly, disturbing and belligerent when someone tries to control him. Acute mania can also produce ideas of self-aggrandizement, delusions of grandeur and paranoia. Patients go on spending sprees out of proportion to their finances. One bought six houses in one week, while another went three days without sleep and claimed to feel fine. A classic example is that of the famous musician and composer, Robert Schumann, who was a manic depressive. During one of his manic phases, he wrote one hundred and thirty songs within four days time.

Delirious mania is a step beyond acute mania. The patient is incoherent and disoriented. Excited and agitated, he wears himself out with pointless activity. It is essential to calm him by physical or chemical means, for he may harm himself or others, and even die from exhaustion.

What is happening in the brain during these phases? Protein is metabolized into the mood-supporting chemical norepinephrine (noradrenalin), and the thinking chemical dopamine. Sometimes both are affected, sometimes only one, so the patient can suffer from either or both a thinking and a mood disturbance. In pure manic depressive illness, the metabolism of norepinephrine is disarranged. When the patient is in high mood, he has too much, in low mood, too little. Neurotransmitters, such as serotonin, which are associated, may also be out of balance.

In short, the brain has lost control of the metabolism of protein, which results in episodes of mania or depression, depending on the level of the body's norepinephrine. The phases do not always alternate, but can run in a series of highs and lows, all lows, or all highs. The reason for this is unknown, but is probably inborn and not due to outside influences such as life events, tensions, a psychological shock, or the attitudes of close relatives. The use of alcohol, narcotics, psychedelic drugs or physical illness will exacerbate these moods.

The main feature of this illness is its recurrence. If a person has a high manic phase, with sleeplessness, disturbance in thinking, and irrational overactivity, it is followed by depression, and the cycle is repeated with alternating periods of depression and elation, or vice versa. There seems to be no relationship between the duration and the time between attacks. There is, however, a relationship between the depth of depression and the height of elation, for the higher the elation, equally deep is the depression that follows.

This illness affects about half of one percent of the population. Thirty percent of cases have but one episode in a lifetime, but most people suffering from more than one episode will continue to have mood swings all life long. Symptoms first appear between the ages of twenty and thirty-five and rarely occur in childhood or after forty. Women are more susceptible than men, comprising about seventy percent of cases.

The treatment can be difficult. It usually takes the manic depressive a year or more of therapy to understand his feelings and anticipate his moods. Most patients enjoy their high moods and usually don't report them to the doctor. They are apt to push on into euphoria through drinking, late hours, and increased sexual and social activities. They won't listen to warnings, they stop the medication, and refuse medical care. During this phase, they spend money like water and behave irrationally. If the patient is a drinker, alcohol creates a destructive syndrome, paranoid overactivity, hostility and even assaultiveness. Drugs will also worsen symptoms and actions.

The personality makes a difference. The thinking person's mood disturbances usually result in paranoid and confused behavior, the doing person's in physical activity that creates untenable situations. The doer is also more apt to compound the problem by the use of alcohol and/or drugs. Both types tend toward sexual excess in the manic phase, more violent in the doing person, and creativity is stimulated in the thinker, but results in work that is flawed.

What is the prognosis? There is complete recovery after each episode. Repeated attacks don't leave irreversible damage. However, death can result from suicide during the depressed phase and manic patients are at risk of heart attack or financial dissipations and disastrous personal interactions. A careful watch should be kept on both phases of the illness.

Because of the erratic progress of the disease, it is difficult to predict when the next attack will take place. Studies show that the more frequent they are, the more often they are apt to occur.

How is this illness treated? The treatment is effective if followed, and is a combination of medication and psychotherapy. Psychological reactions should be discussed with the patient so that he understands the illness, especially in its manic, euphoric stage, which is the most dangerous. If this stage can be controlled, the depressed phase is more easily tolerated.

The importance of treating the mania cannot be overemphasized. This is done with an anti-high mood drug given alone or with one of the phenothiazines described under schizophrenia. The depression is treated with anti-depressant drugs, but may be unnecessary if the mania is properly controlled so that the depressed phase is subdued. Above all, psychotherapy is important. The patient needs support and insight into his welfare, which therapy can provide. It teaches him what is going on in his body, since manic depressive psychosis is actually a physical illness. It is hard for him to realize he is ill during the manic phase, when he feels so well. He rejects help then when he most needs it. He is more likely to call for assistance during the depressed phase and is only too willing to cooperate.

Manic depressive illness is difficult for everyone involved with the patient for at least the first year. It takes an understanding family working closely with the physician to supply the supportive network the patient needs. It is wise to involve the family and let them know how important their role is in the management of the patient, who must be watched carefully but unobtrusively, and to expect and take seriously talk of suicide. Furthermore, it is essential that someone make sure the patient takes his medication, especially in the manic phase, for it is during this time especially that he is unwilling to accept another's opinion. Psychotherapy can change this and teach him to distrust his euphoric feelings. As one patient put it: "Doctor, I've learned that if you say white is black, it is black, and if you say black is white, it is white."

What is the cause of manic depressive psychosis? Genetic origin has long been suspect, and the means to find out for sure now exist. Recent research at the National Institute of Mental Health and other groups strongly suggests that the disease is inherited from a dominant gene. This has been identified, but the evidence is not as straightforward as in Huntington's disease, the single gene of which shows up in every generation, and all who carry it develop it. Could manic depressive psychosis be a recessive gene which skips many generations before appearing again?

The Amish people are examples of this possibility. They live in Lancaster county, Pennsylvania and have a prevalence of manic depressive illness. They are descended from fifty couples who emigrated from Germany in the 1700s and now number around 12,500. They are simple people with strict social mores. They keep to themselves, avoiding all contact with others, have large families and live off the land. While studying the Amish families affected with manic depressive illness, scientists have determined that not everyone who carries the gene develops it. Although it appears in every generation, only one-quarter of the children are manic depressive. Why? Other factors must be involved, such as other genes that have a modifying effect, the environment or a combination of both. The next step is to look for the genetic marker. This is done by using the tools of molecular biology as in the case of Huntington's disease. At this stage, chromosome 11 is suspect, and by eliminating one chromosome after the other, scientists will eventually be able to find the marker.

Recently, new research reveals more clues. A group of scientists from three Eastern medical schools report that they have found a genetic marker for manic depressive psychosis. Concurrently, two other groups have been unable to corroborate this, and claim that there may be more than one gene involved. Based on the finding of the genetic marker and this hypothesis, it is assumed that some cases are caused by a dominant gene on the tip of the short arm of chromosome 11, the tyrosine hydroxylase gene, and in the synthesis of the neurotransmitter dopamine. As only sixty to seventy percent of those who inherit the gene develop the disease, unknown environmental factors may determine those who do.

The groups who could not find a linkage to chromosome 11 did their research on two families in which manic depressive illness appears to be inherited through a dominant gene that appears in every generation. It may be that the genetic causes differ, even though the illness manifests the same symptoms. This premise allows environmental and other factors to enter the picture in cases that do not show the genetic marker on chromosome 11.

These findings lead the way to further research to discover why thirty to forty percent of those with the gene never get the disease and to ways to prevent it. For the first time, molecular genetics is paving the way to new and exciting discoveries about not only manic depression but all mental illnesses.

HOW DO YOU COPE WITH MANIC DEPRESSIVE ILLNESS?

Someone close to you is diagnosed as manic depressive, let's say, your husband. This is a shock, and you don't know how to handle it. It started when he suddenly began doing strange, irrational things, such as buying tickets for a world cruise on the Q.E. II without consulting you.

It may take a year or two before you and your family understand this illness, but your husband can be helped. First, it is important that he begin therapy early, and that you and members of the family do likewise. Remember that he is not responsible for what he does while in a manic phase, and it is especially important that he take the medication during that time, when he is apt to do things he later regrets.

So, when you notice your husband becoming a little "hyper", going a little faster, press the medication, limit the coffee and cocktails, encourage frequent rest periods and don't plan stimulating events. During this phase, he feels so well he doesn't think he needs medication or rest or even sleep. He keeps going every second of the day and much of the night, for the manic patient needs little sleep. Create a calm, quiet atmosphere until the episode passes, and it will.

After a manic episode, he will fall into a depression, deepened by remorse over what he has done. The depth of the depression is gauged by the height of the mania, the higher the mania, the deeper the depression, but medication and psychotherapy can prevent the downward swing. As he calms down, expect him to have a downswing in mood. Medication may avert this, but during depression he needs a great deal of support from you and his family. Encourage him to exercise, invite friends to the house, arrange functions he enjoys and see that he is with people. The antidepressive medicine will be very effective during this time.

Equally as important as medication is proper nutrition, supplemented with vitamins and minerals. Your doctor will explain. Electro-shock therapy is also helpful, but is not used as much because drugs are so effective.

After a while you will be able to anticipate your husband's moods, how high and how low and how often. His moods may swing once a year, monthly, once a week or every day. There are variations in how high patients swing and how low. Some only swing one way, called unipolar. Manic depression is bipolar, both up and down.

Here are examples of the manic phase. A woman came into my office and announced with pride that she had just bought six houses. She considered this perfectly acceptable. Another patient called to tell me that he had just beaten his wife so badly he had to take her to the hospital. I knew him as a gentle and sweet person who loves his wife, but she spoke back and he pommelled her. Another spent thousands of dollars in one day on lavish meals and expensive articles he had no use for. Another drew out of the bank $25,000 to buy a Cadillac, he already had two cars. He handed the cash to the mechanic and told him to deliver to his house. He never saw the car or the money and paid no attention.

Manic depressive patients have poor judgment during the manic phase and little concern, except to get their way. Furthermore, they feel quite comfortable about what they do, which is the antithesis of how they feel when depressed, when everything they do is wrong. Patients complain that their colleagues move too slowly and get in their way, interfering with them and not understanding them. The reaction to this makes the manic hostile and brings out paranoid tendencies which create trouble.

HOW DO YOU, THE PATIENT, COPE?

You went to the doctor because you didn't feel like yourself. First you were depressed, and then developed a "high" and did crazy things. Once, you took the money out of the savings account and flew all over Europe. When you returned, your wife was furious and you went into a deep depression. Then, you both knew something was wrong.

The doctor diagnosed manic-depressive psychosis, and gave you pills to take when you are high and others when you are low. He said that you will have these ups and downs and the higher you go, the deeper will be the depression. He also said it would take a while to come to grips with your illness, perhaps a few years, in order to keep it under control. He warned you that you must be sure to take the medication, especially when you are on a high, for that is when you are apt to do something you will regret. There are countless stories of patients squandering their money and obligating themselves to commitments they cannot afford, only to wake up to find they are grossly in debt. Others commit acts they are later ashamed of, such as indulging in sexual orgies or beating their wives.

Everybody has changes in mood, but yours are more extreme than the average. So, watch yourself, and when you are going a little too fast and feel terribly good, remember that you are working up to a high that can get you into trouble. Try to recognize the signs, slow down, drink less coffee and one cocktail instead of two or more, and stay away from situations that stimulate you. Rest frequently, even though you don't feel the need, and if you can't sleep, have a quiet time alone. During your low moods, try to be with people, indulge yourself in pastimes that excite you and exercise as much as you can. This will give you a lift.

You will learn how your opinion of yourself changes with your mood. When you are high, you can do no wrong, you are right and the world is wrong. When you are low, you hate yourself and want to die. So you have to keep fighting against the tide. Slow down when you go fast, speed up when you go slowly, and the medication will help. When you are high, you need less sleep, some say they need none, when you are low you need more, but going without sleep catches up with you eventually, for sleep is a necessity.

When you are in a high mood your energy is enormous, and you can accomplish much more than in a normal mood, but watch out, for you will find that much is wasted work. You may make a lot of money and sometimes create good things, but you will realize later that much of this energy was expended on useless programs. However, you insist that you are doing the right thing and won't listen to your family and friends who are trying to stop you from making a fool of yourself. For if you are the type who ignores advice and has a fling, your nervous system will burn itself out and you will fall into a depression as deep as your mania was high.

When you are manic, you have poor judgment and are wrapped up in yourself and feel very comfortable about whatever you do. You want your way and will challenge anyone who tries to cross you. On the other hand, when you are depressed, nothing you say, do or think seems right, although to outside observers your performance may appear proper.

So, remember that the most important time to take your medication is when you feel great, for it controls the manic phase and the depression won't be so distressing or so low. When you are depressed, you accept all the help you can get. So take the medication the doctor gives you, for when you hate yourself and just want to crawl into a hole and die, that is only a symptom, not reality. The symptom will vanish with modern treatment, allowing you to live your normal life.

SCHIZOAFFECTIVE DISORDERS

There is another group of disorders called schizoaffective. This is a kind of wastebasket term. They are more difficult to classify as they have both disturbed thinking and swings in mood, so they are a combination of schizophrenia, a thinking psychosis and manic depressive psychosis, a mood disorder. These patients have symptoms of each in fairly equal amounts which differentiates them from the single psychosis of schizophrenia or manic depressive psychosis. The underlying cause of this is also disturbance of protein metabolism, but includes both thinking and mood neurotransmitters.

Is there a chemical difference? Yes, there is an imbalance of dopamine, which produces schizophrenic symptoms, and of norephinephrine and serotonin, causing mood disturbance. All schizophrenics also have some mood disorder, and all manic depressives disturbed thinking, but in lesser amounts. Clear-cut cases are uncommon, but until we know more about brain chemistry, we have to settle for these diagnoses. Eventually, we will be able to classify all schizophrenics and manic depressives as varieties of schizoaffectives, and specify the exact proportion of thinking and mood disturbance in each case.

Treatment is psychotherapy and medication, the same as for schizophrenia and manic depressive illness outlined in those chapters. Acceptance and understanding are the main elements in handling the patient with a schizoaffective disorder. Turn to the chapters on schizophrenia and manic depressive psychosis, where you will find how to cope in greater detail.

Not much is known about schizoaffective disorders, however, it is believed that they may be the key to the total understanding of the psychotic mind and normal mental function. This is a subject about which little is known and may be the cutting edge of research into brain function where the greatest advances will be made.

A schizoaffective disorder combines the thinking and the mood together, which means that there are neurotransmitter defects in both the thinking and the doing areas. Research will show how the combination of thinking, doing, mood and self-management work together and how the thinking affects the doing and the mood affects the thinking and how they all relate together.

In the next ten to fifteen years scientists will come up with these answers which will show the way to improved medication. Today, these patients are treated with a mishmash of antipsychotic drugs, mood control drugs, lithium, tranquilizers, sedatives and sleeping pills, none of which strike directly at the problem.

9 DEPRESSION

There are three kinds of depression. Their symptoms are similar. It can be endogenous, meaning originating from within, reactive, developing from without, and organic, from damage to the brain. We place emphasis in this chapter on endogenous depression because it is the most serious and the least understood by the public.

Endogenous depression (also called chemical depression) is caused by an imbalance of chemicals in the brain. It is a mood disorder and is unipolar, meaning the depression is one way, down. Manic depression, as explained in chapter 8 is bipolar, meaning that the mood swings up as well as down. In unipolar depression the brain's protein metabolism -becomes faulty, allowing a chemical imbalance to occur between the cells of the central nervous system. It is the most common form of depression and can be controlled and in many cases cured. The treatment is with anti-depressant drugs coupled with psychotherapy.

What is reactive depression? Depression may develop as a reaction to the loss of a loved one, a job, an accident or an illness. In other words, after any kind of upsetting life event. This type is sometimes called situational depression as the aftermath of a disastrous life event can leave one in a distressful and troublesome situation. This type of depression calls for psychotherapy as its cause is psychological. Chemical imbalance in the brain is not involved and anti-depressant drugs are not indicated.

What is organic depression? This depression can develop after an injury or damage to the brain from a stroke, an infection, a tumor, or destruction of brain tissue from an accident. It is a physical brain failure, which is what organic means. The different types of organic brain disease are described later under the section THE DAMAGED BRAIN.

So we have three types of depression: endogenous, which is chemical, reactive, which is psychological and organic, which is physical.

Unlike some mental disorders, there are no mysteries about depression. It is well understood and has been researched thoroughly and extensively, resulting in effective techniques for treatment. Ironically, however, recent studies show that up to eighty percent of cases are never treated.

Depression is poorly understood by the average layman, but many people believe they know all about it as most have experienced blue moods personally or seen it in a relative or friend. Much of this experience is interpreted in a distorted manner because of old fashioned psychiatry which describes a depressed person as feeling sorry for himself and therefore needs a kick in the pants to bring him out of it. They don't understand the real cause, that endogenous depression is due to a chemical imbalance which can be corrected by medication. Furthermore, many doctors and even some psychiatrists unfortunately still don't understand the nature and treatment of this type of depression.

Depression is all around us. This became glaringly apparent after an article by columnist Ann Landers on it a few years ago prompted 250,000 requests for more information. Interest is further aroused from exposure by the press of the alarming increase of teenage suicides. Suicide is always preceded by severe depression, often triggered by abuse of drugs or alcohol. In addition, there are more articles on depression, both in the scientific and the lay literature, but this does not seem to have had sufficient impact. This is due to the vast differences between the various forms of depression. Lay persons lump them into one and therefore feel that one treatment is correct for all.

People are confused by the word depression and some think of it as in the cartoon "Charlie Brown". Charlie says: "I feel depressed" when something happens that upsets him. This is reactive depression and is because of a distressing event. Any pressure, depressed anger or an internal psychological factor can cause it and it will pass. Endogenous depression is not caused by any of these, but can be aggravated by any of them, and the depression deepened or the treatment made more complicated. So, endogenous or chemical depression is accurate but its description is difficult for the lay person to comprehend.

Perhaps we need a new name. What shall we call it? How about an "energy illness"? That is what endogenous depression actually is, the loss of mental energy which reflects both on the mind, causing a loss of intellectual energy, and on the body, causing a loss of physical energy. What happens when energy is down? You feel depressed and your physical energy is very low, you are confused, agitated, irritable, constipated and have difficulty sleeping. Your concentration is poor and your sex drive is diminished, you feel guilty, hopeless and ill and are afraid you are going insane. In addition, you lose your appetite and physical energy, depending on the severity of the loss of brain energy. But why can't you sleep when you are so tired? Because brain rhythms that create sleep need energy to function.

It is difficult for a doctor without psychiatric training to diagnose endogenous depression. Usually, the patient complains of physical symptoms and seldom mentions the changes in mental outlook, because these are overshadowed by physical complaints. This is misleading, and a medical doctor may assume a physical condition is causing the symptoms and the correct diagnosis is missed.

What are the symptoms? They are physical or psychological or both, and can come on simultaneously. A loss of drive and lack of initiative may be the first indication that something is wrong. You have to force yourself to do the simplest tasks, and your regular activities seem overwhelming. This is a major change from your usual self. You have always been able to cope and now you have doubts that you can. You dread the responsibilities you have shouldered and devise means of avoiding them, at the same time feeling deeply ashamed of yourself. The future looks foreboding and you feel threatened by it. This terrifies you and the fear of becoming insane is always present. This never happens.

Tears are seconds away as your depression deepens. Nothing matters. You hate yourself and can't stop the horrible images that come to mind. If you could only stop thinking! You want to crawl into a hole and die, for death seems the only release from the torture. It is agony to meet others and you avoid all contacts, even your family. Loss of concentration is the most telling symptom and is always disabling.

Most patients think of suicide, for whenever brain energy is low the desire to die is triggered. "What is the point of living?" "What use am I?" "I am a burden to my family and myself." "I am better off dead." These are natural thoughts and feelings, not signs of impending insanity, and are symptoms of the illness. Although treatment and cure is quick and complete, it is difficult to believe that you will be well again, so you need closely supervised support during this time, and it is wise to seek medical help. Treatment is available and simple, and it works.

These episodes are cyclic and self-healing and very seldom end in suicide. Endogenous depression is self limiting and disappears as fast as it appears, even without treatment. So, if you wait long enough you will be well. However, it may be months before the mood shifts naturally, so treatment is advised. As soon as you get treatment you only have to suffer these agonizing symptoms for as long as the medication takes to be effective, usually two to three weeks.

Hopelessness is another symptom of endogenous depression and is at its worst in the morning, better at noon and even better at night, but it is not a constant, overwhelming force. It fluctuates, and some days you feel pretty well, but most days you are miserable, especially in the morning. "What am I good for?" "How can I work?" "Life isn't worth living." "Can't somebody help me?" Yes, your doctor can.

You also have physical problems which upset you. These are called "vegetative" signs and may be poor appetite, loss of weight, insomnia or over sleeping, nausea and vomiting or constipation, chronic fatigue, listlessness and vague aches and pains. Sexual desire is greatly reduced. You don't feel well and don't know why.

Emotional stress uses up brain energy, so you fall deeper into depression as anxiety and disturbing thoughts envelop your mind. Your body is working on your mind and your mind on your body, so the brain chemistry becomes more unbalanced. What is the cause of this and why? Every family has a history of depression. It may be generations back, but susceptibility is always inherited. Who succumbs? Endogenous depression can occur at any age, usually in the thirties and forties, and high energy people are the most affected, men and women equally. The incidence is high, roughly three percent or over seven million Americans.

What happens in the brain? The chemical norepinephrine is the neurotransmitter that shuttles between cells carrying messages. It goes back and forth being secreted and then reabsorbed by the brain cells. A failure of production or too much reabsorption reduces the norepinephrine to an ineffective level. One theory is that not enough is secreted, while another that too much is reabsorbed. Serotonin, another transmitter, may also be involved, and whether it is faulty in secretion or reabsorption too is not understood. Other protein substances are released also by the brain cells, but their function is as yet unknown. The search for the gene responsible for endogenous chemical depression is under way.

The term depression is actually weak and inadequate for a myriad of emotional and physical conditions. As it comes in many forms, doctors are presented with a difficult diagnostic challenge and most psychiatrists have the ability to identify endogenous, meaning chemical depression. Once this is done, they can tailor-make the treatment and prescribe the medication for each particular patient. This type of depression is effectively treated with antidepressant drugs and, if psychotherapy is administered with the medication, the results are always better. The real difficulty is in making the diagnosis without an available test.

What can you expect if you visit a psychiatrist for a diagnostic session? He will evaluate your mood by several indicators, such as the answers to his questions about your history, your symptoms, your diet and your family. He will want to know what drugs you take and your use of alcohol, and he will inquire about your sleep and work habits. He also observes your manner, dress and personal appearance.

Once the diagnosis is made and the doctor is convinced that the depression is due to endogenous causes, he can apply psychotherapy and prescribe antidepressant medication. This last is the easy part, and within two to three weeks the drug will take effect and the patient will feel much better, the mood will rise, the appetite improve, the drive and will to live return, and the physical symptoms disappear.

So the difficult part about depression is determining the type of depression and making the correct diagnosis. This ability only comes through training and experience, as not every doctor can differentiate between a neurotic depression, a chemical depression and a reactive or situational depression. This takes psychiatric training which most doctors do not have.

Isn't there any kind of diagnostic test as there is for so many other conditions? Yes, there is, but only with dexamethasone, and this proves to be inconclusive in a high percentage of cases. However, there may be new hope. Scientists in this country and in Germany are experimenting with a combination of dexamethasone and another biochemical assay and this appears to improve significantly the diagnosis of endogenous depression.

Reactive depression can worsen endogenous depression. If a traumatic life event is superimposed, like an illness, an injury or the loss of a loved one, the depression is compounded. One type feeds on the other. A recurrence of endogenous depression is triggered more often by a physical than an emotional shock, reactive depression by an emotional stress. Treatment of emotional trauma is psychotherapy, while endogenous depression requires chemical treatment with supportive psychotherapy.

Most cases of endogenous depression come in cycles and each person's cycle is individual. In some it is daily, some weekly, some monthly, six months or yearly. The cycle is irregular and unpredictable. Some people are free for months or years, only to have it recur unheralded. Seventy-five percent have only one attack, while twenty-five percent experience recurrences. Endogenous depression can also be triggered by the abuse of alcohol or drugs.

It is a condition that comes and goes, with or without treatment, and each episode is of uncertain duration. Indefiniteness causes fear and anxiety, but knowing that treatment is sure, safe and close at hand should dispel fear of the unknown.

Years of observation show that the movement of the earth around the sun has an effect on the cycle of endogenous depression. The months of February and August are particularly notable, with May and November next. More patients suffer a recurrence in these months than any other time of the year. Why? Could it be due to the earth/sun relationship when the angle of the rays is changing in February and August? The amount of light from the sun is equal during those months. Perhaps it is another happening in the cosmos we have not yet discovered. In any event, the exact cause is unknown. Studies show that the occurrence of endogenous depression is not related to the barometric pressure or other natural meteorologic effects, and research at high altitudes indicate that there are no fluctuations that could be responsible in the spring and the fall. Sun spots and gravity waves are not the cause. But in February, the sap begins to flow and the leaves get ready to sprout. In northern climes, even with snow on the ground, house plants shed old leaves, grow new ones and start to bloom, and in August the greenery begins to dry up and there is no new growth. Such natural phenomena reveal a deep, pervasive influence that affects man as well as flora and fauna.

Many people become depressed during the holidays and believe that the weeks around Christmas are apt to bring on endogenous depression. This is not so. This is a reactive depression from pressure, fatigue, and memories of loved ones not present. The intensity of holiday activities, the late hours partying, drinking and eating too much create harmful reactions in the body. Regular schedules are disrupted, people are traveling, children are home, families get together. All this brings on physical and mental exhaustion with the attendant emotional letdown that causes reactive depression.

How do you treat endogenous depression? Once a correct diagnosis is made, the treatment is drugs and psychotherapy. It is imperative to start treatment immediately because of the threat of suicide and the great distress this illness brings. The drugs most used are the tricyclics and the monoamine oxidase inhibitors, combined with psychotherapy. There are patients who respond to treatment in a day or two, while others take two weeks or more. Every case is individual, but all are helped and nearly all are cured permanently. Recurrences are treated the same as the first episode and anti-depressants are safe and effective for those with repeated illnesses.

For the occasional patient who does not improve with drugs and psychotherapy, electro-shock therapy is indicated. One series of treatments, (ten to fifteen), is usually sufficient, and most patients respond and stay well with this technique. Modern improvements have made electro-shock safe and without lasting side effects.

We don't know how this therapy works. Perhaps it allows important nutrients to permeate the blood/brain barrier and supply the brain with the chemicals that have gone out of balance, or the electro-shock stimulates the brain to begin producing again its chemical energizers. However, electro-shock remains the single most effective treatment for endogenous depression. All other therapies are measured against it for comparable results. When no drug treatment and/or psychotherapy helps, electro-shock will remove the depressive illness almost without exception.

We all have biorhythms, and some investigators find that if these are interfered with, depression is relieved temporarily. As an example, staying awake all night, called sleep deprivation, sometimes miraculously cures endogenous depression. We don't know why, but this suggests that melatonin, which is secreted by the pineal gland, may be involved. This may also be linked to "seasonal affective disorder", SAD, which afflicts some people prone to depression only in the winter and are relieved by spending several of the dark hours of each day in a strong light. More about this later.

In 1950-51 research brought hope. The discovery of the drugs caused a revolution in the treatment and acceptance of endogenous depression by the medical profession, patients, and their families. The illness lost its fearful and mysterious quality, and now can be diagnosed and treated easily and quickly.

In the United States alone, fifteen million people are affected by severe depression of one kind or another in a given year. There are over thirty thousand suicides every year, many of young people, and depression is associated with most. Although there are many antidepressant drugs for the treatment of depression, they present serious drawbacks, so it is no wonder that the National Institute of Mental Health is concentrating on finding the cause of depression.

Ongoing research indicates that scientists are in process of decoding the biological basis of depression and are coming close to whether the different types have different causes. At a recent conference on the disorder, the consensus after two days of intensive discussion was that all types of depression share a common chemical pathway which, when identified, can direct the way toward more effective drugs, regardless of how the disorder was triggered, by severe stress, an unhappy life event, a medical or a mental illness, or out of the blue. In the final analysis, the key to success in the treatment of depression will be to match each patient to the drug that is right for him or her.

A famous case of endogenous depression is Sir Winston Churchill, who suffered periodic attacks. He would remark: "The black dog has come again to rest on my doorstep". In his day, there were no drugs, no tricyclics or monoamine oxidase inhibitors. Electro-shock therapy and psychotherapy were the only methods of treatment, yet Churchill pursued an outstanding political career with many ups and downs, and his magnificent statesmanship did more to rescue the western world than that of any other individual.

This is another case. Malcolm was frightened and bewildered when he went for medical care and complained of insomnia and inability to work. He had lost his drive, couldn't concentrate, and was afraid he was losing his mind. He had lost weight, had no appetite nor interest in sex. An antidepressant drug, a mild sleeping pill and psychotherapy resulted in relief in two weeks. Medication was stopped after six months and to this day he remains well.

Another case is Angela, who experienced disturbing symptoms. She was losing her drive, which had always been high, and had frequent bouts of nausea and vomiting which was foreign to her good health. Several doctors, including an allergist, could find nothing wrong.

Her problems persisted for about a year when an emotional event occurred. Her only daughter had planned with her mother's consent to leave home when she was twenty-one. When the day came to move out, Angela broke down. Then she knew she needed help and went to a psychiatrist who diagnosed endogenous depression. He said: "Take these pills and you will feel better in two weeks." And she did. The medication as a monoamine oxidase inhibitor. However, this was not the end. For the next few years, the depression returned periodically, the same treatment was given and within two weeks time Angela was back to feeling herself.

These episodes recurred every few months and were at times extremely severe, so severe that hospitalization was once required. In fact, Angela asked for it. She said: "Doctor, please give me something to stop me from thinking! I hate myself and feel guilty that I can't work or carry out my responsibilities. I don't want to answer the telephone or talk to anyone, even my children. It is agony to get myself to the mailbox to pick up the mail. I run out of the house and back as though I were being chased by demons. This morning I stood in front of the mirror and tore at my face with my fingernails, obsessed by self-hatred. Then I called you because I was afraid of what I might do. You put me in the hospital and knocked me out. It was heaven."

This was the worst episode, and from than on they became less and less severe and frequent. She went faithfully for psychotherapy and the doctor kept saying that the drug is curative and one day she would be cured, and he was always available over the telephone for support. Fifteen years after the first episode, there were no more.

THERE ARE VARIATIONS OF ENDOGENOUS DEPRESSION

Seasonal affective disorder, referred to as SAD, is the technical term for winter depression. As the days shorten in the fall and standard time takes over, further shortening the daylight hours, a group of susceptible people slow down and sleep more than usual. Some eat more and all gradually become depressed, resulting in an inability to function as formerly.

Studies of this group in several laboratories show that by lengthening the daylight hours through artificial lighting in the late afternoon, winter depression is reversed. Experiments were also made supplementing lights in the morning as well. Results showed that one half to two hours of early morning daylight proved to be the most effective. The brightness of the light did not seem to be a factor.

What made this happen? Our daily biological rhythms are delayed and reset by morning light, using the hormone melatonin as a marker for light sensitivity and body rhythms. Melatonin secretion begins when it gets dark, gradually reaches a peak during the night, and falls with the approach of daylight. In these people, it was found that the pattern begins and ends several hours later than normal. After four weeks of light treatment they were all improved.

What does this tell us? It seems that norephinephrine and noradrenalin levels alone do not account for all depressive illness, and that the pineal gland, a pea sized gland located deep in the brain, and its secretions may be aligned in some indirect way. It is well established that depression is more prevalent during the times of the year when there is less daylight, which may touch off depression in susceptible individuals.

Another form of depression called involutional is also a mood disorder of protein metabolism. It can occur at any time after middle life. "Involutional" refers to the change of life period. It appears during the fourth decade in women and the fifth in men, associated with the menopause and the male climacteric. Three times as many women as men are affected. It is an uncommon type, but is identical in nature to the other endogenous depressions with a lack of noradrenalin.

What are the signs? Patients are anxious, agitated and have delusions. A neurosis is suspected but severe depression follows shortly, clinching the diagnosis. Other symptoms are insomnia, loss of appetite, gastrointestinal complaints and chronic fatigue. Depression may bring on suicidal attempts, as all victims feel hopeless.

We can only speculate what the causes are, for the answers are not clear. Changes in the hormonal balance at the time of the menopause are suspect. We don't know what sparks the onset, or what connection, if any, there is between the hormonal changes and the shift from normal protein metabolism to imbalance.

Although most of these patients have obsessive compulsive (prepsychotic) personalities, why they develop involutional depression and the majority doesn't, is an enigma. Environmental factors may be responsible, but are without proven foundation, and there is no known measure to prevent it. The initial therapy is electroshock, which brings the patient quickly out of depression. This may be temporary, but usually is permanent. Anti-depressants and anti-confusional agents are also effective and estrogen therapy is recommended for women going through the menopause, but is not a cure in itself. Psychotherapy belongs with these treatments and is the cement that binds the chemical cure. About ninety percent of patients respond favorably to treatment and return to regular activities.

HOW DO YOU COPE WITH DEPRESSION

Whether reactive, involutional or endogenous depression, the symptoms are the same. They all vary only in intensity. Endogenous depression is the most serious.

All depressive states require acceptance and understanding, without criticism or judgment, and it is difficult for one who has not experienced depression to know what it is like. If you are the patient, be assured that it will pass even though you hold no hope.

> Remember to have confidence in your doctor and believe what he says and do what he tells you.
> You will feel yourself again.
> Your horrible thoughts will disappear.
> Your life will return to normal.
> Be patient and accept the treatment.
> Exercise as much as you can.
> Take the medication prescribed by your doctor.
> Do not force yourself to work.

If you are a spouse or close to the patient, give in to his wishes. He may not want to see much of you, if at all. He may not want to go to work, or out of the house or talk on the telephone. Some depressives are suicidal, and this should be carefully watched and any attempt reported to the doctor.

It is hard to understand why the patient blames himself for everything, and there are times when he puts pressure on you by insisting that he is making you as miserable as he feels. If you say: "No, I feel fine", he doesn't believe you and insists it is all his fault that you feel that way and he would like to kill himself. This is sort of reverse reasoning, projecting his illness on you because he cannot imagine anyone feeling any other way.

It is well to monitor the patient's medication and see that he goes for therapy, for this will bring him back to normal. Exercise is excellent for raising the mood, and your patient should be encouraged to indulge in whatever form he enjoys as often as possible. And remember that the illness will pass in time.

10 OBSESSIVE COMPULSIVE DISORDER

Obsessive compulsive disorder is a thinking and doing neurosis involving thoughts and actions. The severity of the illness ranges from very mild to severe. It can begin in a child as young as three but usually does not appear until adolescence or early adulthood. OCD afflicts both males and females and is considered relatively uncommon, probably because when mild it is not considered worth reporting. Today, however, mild forms are found to be fairly widespread.

The symptoms are compulsions and obsessions, the compulsions emanating from the doing and the obsessions from the thinking. Some people suffer from compulsions, others from obsessions and many from both. Each case is individual in the intensity, ranging from mild and unnoticeable to severe and threatening to daily life-style. Even mild symptoms can interfere in a normal life-style.

The compulsions are for perfectionism and are overwhelming. They embody strict responsibility for every thought and act, conscientious scrupulousness and rigid adherence to every principle of the highest standard. The patient with this mental disorder is filled with recurrent thoughts and is unable to refrain from exercising compulsive activity in response.

The obsessions are recurrent, disquieting thoughts, such as a parent fearing he might harm his child or a loved one, or a religious person harboring a compelling concern that he could burn the church down, all the while realizing that this horrible thought is just a figment of the imagination.

The compulsions are impulses to do ordinary acts many times over, such as washing the hands, checking to see that doors are locked and lights and appliances turned off, counting the stairs or cracks in the sidewalk, counting cars going by, repeating a question many times, and so forth.

These obsessions and compulsions cause marked distress and interfere with daily life, consuming countless hours of senseless activity, the aimlessness of which is quite apparent to the individual, thereby adding to his discomfort. Coupled with this is the inability to curb these irresistible drives without giving in to them.

A patient senses that he can catch a lethal disease from an object he cannot avoid touching, such as the keys to his car, and washes his hands repeatedly afterwards. Or he checks to see if a door is locked over and over or runs into the kitchen three or four times to make sure the stove is turned off. These actions are so time consuming that ordinary activities are disrupted and the illness becomes disabling. And if the tension becomes more than can be tolerated, he gives in and obtains immediate relief.

So this is what obsessive compulsive disorder is about, and when asked how much of the day these rituals absorb, patients reply "All day long."

Just what is this strange behavior all about? Obsessive compulsive disorder is an exaggeration of a normal brain function, just as schizophrenia is an exaggeration of a normal thinking process and depression of a normal mood process. Therefore, these mental illness are extremes of what goes on normally in the brain. Compare obsessive compulsiveness to practices we had to learn as children, such as buttoning buttons, tying shoe laces and bows, putting on a coat and a sweater, brushing hair and teeth. Once we mastered them the know-how stayed with us as obsessive compulsive acts. We don't have to think about them now, they just take place naturally. They are not habits, but long term memory muscle motions that are locked in for life. It is the same with swimming, typewriting, skating, playing the piano and so forth. We don't have to relearn them every time we use them. They are there to be drawn upon whenever we want to. So we are created in such a way that we can learn to use the muscles that enable us to do certain things and these abilities become implanted in our memories for life, linking body and mind and making life easier. If we had to relearn them every time we wanted to use them our lives would be in shambles.

Although these functions are close to habits, they are not, for habits can be broken and memory muscle behavior cannot. Watch a baby learn to eat with a spoon. He begins by grabbing the food with his hands and pushing it into his mouth. This is a primitive response and a method of assuaging his hunger. When a spoon is placed in his hand and guided to his mouth, he tries to do it himself and, after a few days becomes fairly adept. It is quite a struggle, but when the feat is finally conquered, he will repeat the process millions of times throughout his life without a thought as to how he does it. It is a muscle function that takes place automatically.

One of the most common obsessions these patients contend with is the fear of catching a disease from food that is cooked by someone unfamiliar, or touching a person or any object another has come in contact with. In other words, an obsession about germs and/or dirt. However, a person so afflicted is very aware of his obsessions and accepts that they are phantoms of the mind which have become uncontrollable. So the intelligence rebels and he says to himself, "Stop this nonsense!". This only results in increased tension relieved by giving in. At times, anxiety becomes so oppressive that the individual has to give in to the obsession and when he does, the pressure is immediately relieved. Then the vicious cycle continues. Prolonged stress from this recurring scenario may result in depression and bring the patient to a doctor.

Here are examples of obsessive compulsive disorder. Bernardino was the wife of the president of a large corporation and suffered from COD. Luckily, the family's affluence allowed her to indulge in her neurosis and a host of servants catered to her and acceded to her every wish. Her major problem was fear of dirt. She dressed only in white and kept a pile of white towels nearby with which she touched objects she had to, such as doorknobs and light switches. She had boxes of white gloves which she used once, then discarded. She ate only what one special servant prepared for her and used gloves to hold the tableware. When she sat at the dinner table, her chair must be covered with a white sheet and when she goes out in her chauffeur driven limousine, the driver must wear white gloves and spread sheets over the seat she occupies. This is also true for the theatre she attends, where three seats are reserved and draped with white sheets.

Another case is about three year old Jamie. When he crosses the street with his mother, he is compelled to circle the manhole covers and in kindergarten he sits for hours making circles on paper. At eight, he stands up and sits down seventeen times before finally sitting down. Belinda was twenty when she began to pull her hair out strand by strand until she was bald. The hair grew in as she pulled it out so it was a continuous process.

Catherine could not leave the house until she checked over and over again to make sure the appliances were off. This ritual usually took about two hours before she was on her way. During one day, Mary washed her hands forty times, counting how long it took her each time.

However, sometimes, the disorder is less debilitating, as in the case of sixteen year old Tabitha. Every Sunday, she removed everything from her room and scrubbed the walls and floor. Otherwise, her life was that of a normal teenager.

A well known sufferer of obsessive compulsive disorder was Howard Hughes. As a young man, his fear of germs began and slowly increased as he aged until it became so excessive that he would touch nothing without a clean pair of gloves and eat only what a particular servant prepared for him. His compulsions consisted of repetitive sentences and orders to his staff which consumed a good part of every day. He began to deteriorate seriously when he took to hard drugs. He died a recluse.

Do we know the cause of this disorder? We don't know the exact cause, but researchers believe it is biochemical in origin and heredity may play some part, as about twenty percent of patients with OCD have family members with the same disorder. Today's sophisticated imaging techniques show minor abnormalities in the brains of these patients, especially in the frontal lobe which houses activity in regard to fastidiousness and meticulousness, and the brains of OCD patients demonstrate overactivity.

Originally, it was believed that the cause comes from disturbing experiences and unresolved conflicts of childhood, but in recent years, the focus of research is on a derangement of function and/or a chemical imbalance in the brain similar to depression and schizophrenia. What led to this was the observation that symptoms of the disorder improved when key regions of the brain were partially destroyed by surgery or damaged by accident or disease.

Recently, three groups of scientists have come to the same conclusion, namely, that the brain metabolism of these patients is excessively fast and burns up an abnormal amount of energy. This can be seen by PET scans, which are radioactive studies of brain metabolism, and show that certain parts of the brain burn up sugar at a much faster rate than in normal persons. The focus of these studies are the basal ganglia deep in the brain, particularly the actuate nuclei. Glucose, or sugar, is the brain's source of energy and there is an increased energy use in these patients. Just how and if this causes the symptoms of the disorder is unclear, but it gives a clue toward the line of research to pursue.

Until recently, research was focused on the patient's obsession, for example, constant hand washing. In the last several years, however, the content of the obsession has not been considered important, and the focus is on understanding the patient's vulnerability to repetitive behavior. The parts of the brain highlighted in the PET scan studies are believed to be decisive in the control of repetitions. For example, these parts may tell the brain not to worry about contamination once the hands are washed.

One scientist notes increasing evidence of chemical abnormalities in the brains of these patients, and believes that OCD may be more than a single condition. Recent studies indicate that many patients have excessive amounts of the neurotransmitter serotonin in their brains or are abnormally sensitive to it and that drug treatment increases its level. Whereas it was once thought that symptoms were caused by a shortage of serotonin. Abnormalities of other neurotransmitters are also found to be involved in this behavior and are currently under study.

A composite picture of the new research offers clues to the kind of chemistry that is abnormal in the brains of these patients and where it is detectable. It is still unclear, however, how much represents causative factors and how much results from the disease process. The brain controls all physical behavior and mental activity and depends on internal biochemical processes. This research gives leads to the sites in the brain that may be most involved in the abnormal thinking and behavior of patients with obsessive compulsive disorder. As one scientist put it: "We are so much closer, yet so far away."

Can this disorder be treated? Yes. More than sixty percent of patients respond favorably to drug therapy. So far the most effective drugs are Prozac, anafranil and clomipramine. Yohimbine is indicated when sexual dysfunction is reported, which can be caused by other drugs as well. Psychotherapy is not effective and is used as an adjunct for the support of the patient.

The course of OCD is chronic, waxing and waning over the years. Drugs are more effective when accompanied by behavior therapy for the depression that may ensue, however, drugs are not conclusive in dispelling the disorder. Tranquilizers may be useful temporarily and studies on more drugs are under way.

Behavior therapy is demanding of the patient and about twenty-five percent do not choose to continue, however, it is recommended that this be tried first, and if not effective to try medication. It appears that some patients respond best to drugs and others to behavior therapy. The bottom line is that the results are excellent if the person sticks to some kind of treatment.

At this time, our knowledge of the treatment of OCD is considered successful if those who are treated no longer consider that it interferes with their lives, even though they may still have vestiges of symptoms. They may check once or twice to see if the door is locked and the keys to the car are in the house, but don't repeat these procedures over and over.

11 ANOREXIA NERVOSA AND BULIMIA NERVOSA

Anorexia nervosa is a disorder resulting from a defect or disturbance of the appetite regulatory center, a tiny section in the hypothalamus located in the mid-brain. This center controls the sensations of hunger and of being satisfied, and is also connected to the brain's self-concept area, so the disturbance in self-image goes along with the disturbance in appetite. Actually, anorexia is a misnomer, for it means "no appetite" and these patients are extremely hungry. The disorder occurs in one in two hundred women aged twelve to thirty-five. About one third are obese to begin with.

This eating disorder has psychological and biological components with severe voluntary restriction of food resulting in emaciation, loss of menses, infertility and sometimes death. There is a block somewhere along the maturing process. The teenager starves herself, and this creates the eating disorder, whereas the older woman develops a mental disorder and then becomes anorectic. So disturbed mental function comes first and is the key to the eating disorder in the older woman, and the eating disorder affects the mind and results in disturbed mental function in the younger girls.

What are the symptoms? The anorectic simply will not eat and loses weight until she is skin and bones. Remarkably, she is very active in sports and exercises strenuously without fatigue, and when asked why she does so, claims she is overweight, or wants to become a model, or just is not hungry. Efforts by her family to make her eat are resisted by every means, and although she eats at times, she makes herself vomit soon after. At a critical point of weight loss, her menses cease. When she looks in the mirror, her self-image is distorted, for although she is painfully thin, she sees a plump girl, which reinforces her determination not to eat. Only her own image is warped, for when she sees another anorectic, she remarks how thin the girl is, although she is as emaciated herself.

When told that she will become ill if she does not eat, she insists that she is perfectly all right and feels fine. She may explain that she eats enough and keeps food in her room to munch on. At times, she may go to great lengths to prepare an elaborate meal for the family and, when the meal is served, fill her plate, nibble on it and throw the rest away. If she thinks she has eaten too much, although it may be only a spoonful, she will induce vomiting.

And that is not all. For reasons unknown, the disorder travels to other parts of the brain and causes a breakdown in relationships with women, especially the patient's mother and sisters. Relations with the father, brothers and other men are not disturbed. Battles with her mother and sisters are followed by waves of depression, and when a well-meaning parent tries to force food on her, a violent scene may ensue. Apparently, the patient is fighting underlying depression with rage, and she benefits by creating anger in her family because it helps to overcome the depression that interferes with her life.

Only a small section of the brain is affected, however, for these young girls lead the usual life of a teen-ager. The intellect is intact, as indicated in this case. A sixteen year old anorectic girl ran screaming down the hall of the hospital intending to throw herself down the stairs. The nurses brought her back to her room in hysterics just as her mother came to visit. She quickly calmed down and chatted with her for a while, then challenged her to a game of scrabble, which she won.

How does this start? After an uneventful childhood, the disorder develops rapidly during adolescence and occasionally before puberty. A girl begins to lose weight dramatically and, at the same time becomes very interested in food, which she hoards and wastes. She studies diets and calorie counts but, when she becomes painfully thin, she denies illness and continues to starve herself. A few patients carry this to the point of death, but this can be averted through family support and skillful management by the physician.

What is the treatment? There is no specific curative therapy, but weight must be normalized. Every form of treatment has been tried, electroshock, psychotherapy, and all kinds of drugs, but none is significantly effective. At one time, it was theorized that anorexia nervosa is tied to a girl's fear of pregnancy, but treatment based on this worsened the condition. Psychotherapy helps the patient eat enough to maintain a weight below what it should be, but above what she thinks ideal, and is helpful for the parents, who usually feel guilty when understanding the condition.

Severe cases are hospitalized until weight is stabilized, then sent home under medical and psychiatric care. Mild cases are treated at home. Some patients improve faster if they attend a special school under supervision, where a close watch can be kept, for occasionally suicidal and self-destructive tendencies occur, including drug overdose triggered by depression. Drugs are of little use, but in severe cases a combination of phenothiazine and amphetamine gives support and stimulates appetite, but drugs should not be prescribed for more than a few months to a year.

What is the course of the disease? It shows up spontaneously in normal individuals around puberty, as the sexual hormones collide with the brain center for appetite. As the disease progresses, it initiates a series of complications. The menstrual cycle disappears and at times patients show psychotic signs in self-image and cannot see themselves as they really are. This disorder is not a psychosis although these patients have no realization of what they are doing to themselves by refusing to eat. Severe weight loss ensues, under nutrition triggers emotional and physical symptoms. They lose insight and become vicious, angry, hostile, paranoid, suicidal and depressed.

Most cases of anorexia nervosa automatically reverse themselves, however, it can become a chronic condition, A large percentage go into mild weight gain, develop episodes of depression, and withdraw from family and friends. Some become obese, or develop bulimia, a related condition, or over-eat during an anorectic phase, then induce vomiting. No matter how thin they are, anorectics fight to lose a few pounds. Most patients can be brought to normal weight, if not to normal eating habits. However, the relapse rate is about seventy-five percent.

The cause of this disorder is not clear and there are conflicting opinions, but anorexia nervosa appears to be a developmental disorder sparked by a desire to lose weight, for which several factors maybe involved. These can be a situational, environmental or peer group pressure, and the persistence of the fashion that "thin is in", and overweight is socially unacceptable. It develops from dieting, and eighty percent of patients have unrealistic images of their weight. Young teenagers at risk may diet because of anxieties associated with sexual maturity and social responsibilities as they emerge from childhood. Another group diets in order to attain low weight for professional reasons, such as ballet dancers, models, jockeys and so forth.

Families may play a role in the incidence of anorexia nervosa, as many show an unusually high concern for how they appear to others and how they present their image to society. Furthermore, it is inferred that anorectic girls are prone to strong paranoid tendencies, as they can be very manipulative and consistently strive for power over family members. Susceptible girls may also be conditioned by the dinner table which is when the family is together and conflicts often arise, causing anxiety associated with meals.

Anorectics are bright, capable and energetic young women from the middle or upper classes. The disorder rarely occurs in blue collar or working class families, but recently reports show that there is a higher incidence among poor and American black women for the first time, so the disease is increasing in Western societies. However, it is very rare where there is a food shortage.

Observation over years of experience indicate that the majority of anorectics have alcoholic fathers, and mothers who are borderline between neurosis and psychosis. They are sensitive, emotionally unstable and ineffective mothers. Therefore, there appears to be a genetic predisposition and/or an environmental influence. However, we maintain that a young woman cannot be made anorectic or bulimic by her parents or anyone else.

BULIMIA NERVOSA

Bulimia is an eating disorder that occurs alone or in conjunction with anorexia nervosa. It is a sign that the control mechanism of the appetite is failing. It is estimated that in the United States up to four percent of young women suffer from anorexia and/or bulimia. The bulimic overeats and then induces vomiting, but she never tries to lose weight. She eats to satisfy a tremendous appetite, then makes herself sick, only to repeat the process a few hours later. Such a binge usually ends because of considerable discomfort in the abdomen or persuasion by the family, so vomiting is induced which relieves the pain but may start up the eating again. Although these binges may be pleasurable to the patient, they often result in self-criticism, then depression.

Bulimics are often concerned about their weight and make attempts to control it by various means, such as diet, vomiting or cathartics. So the weight goes up and down depending on whether they are binging or fasting. Many bulimics have their lives dominated by matters of eating, weight control and obsessive focus on themselves.

Again, the cause of bulimia is not clear. One group claims it is a psychiatric problem, another a biological, perhaps hormonal. Recent research, however, indicates that patients with bulimia fail to secrete enough of a hormone that induces a sense of fullness when eating. The study reveals impaired secretion of the hormone cholecystokinin (CCK) in bulimics, which produces the feeling of being satisfied after a meal. This hormone is secreted in the small intestine and is also involved in certain abnormal behaviors, including depression. So it appears that a biochemical malfunction is present and it is possible that a drug can be developed to bring it into balance. This demonstrates that both theories of the cause of bulimia are correct.

What is the treatment? At this time, antidepressant drugs are the treatment of choice, but are not effective in all cases. Psychotherapy helps the bulimic patient not to induce vomiting, and to abstain from stuffing herself in order to reach and maintain her ideal weight. So, as the disorder appears to have both psychological and biological components, psychotherapy and anti-depressant drugs are the accepted treatment.

Studies show that bulimics with normal weight are never sated. They can always eat more, indicating a malfunction in the brain, whereas anorectics have practically no appetite and are never hungry, or are confused over hunger and satiety. Pleasure in taste and smell disappears quickly, accompanied by a feeling of fullness. Bulimics, on the other hand, feel pleasure in eating, which increases more and more as the meal progresses, and they may eat more at the end of a meal than at the beginning.

Dopamine, norephinephrine and serotonin are neurotransmitters in the brain, and play a role in satiety. It is suggested that eating increases dopamine in a section of the brain where a pleasurable sensation rewards us for eating and also tells us when we have had enough. Amphetamine, which releases dopamine, causes the same pleasurable sensation and is addictive, therefore, it is speculated that people who abuse food may be seeking the same reward drug abusers seek. Thus, eating disorders may have characteristics in common with drug addiction. Only research can answer this question.

How do you cope with a child who has anorexia and/or bulimia? The parents and family are under tremendous pressure, so it is mandatory that they understand that the condition is temporary and that they should submit to therapy as well. This helps them understand what the doctor is doing for the patient, how to handle her and what to expect. If the doctor recommends hospitalization, go along with it. If not, the family should be part of the treatment.

However, a young person rarely needs hospitalization, although about twenty-five percent may in order to restore health because they went for medical advice so late, but most can be treated on an out-patient basis. One of the most disturbing signs is when the patient turns against the women in the family, and goes out of her way to antagonize them. Her hostility is almost psychotic in its unfairness and unreasonableness, and her anger creates anger in those she is hostile to. One feeds on the other out of all proportion.

These conditions are serious but not life-threatening, however, it requires continuous understanding and support, and most importantly, cooperation with the doctor's recommendations.

How do you, the patient, cope with anorexia? This is very difficult for you because something has changed within your brain, and you don't feel or think as you used to. Now, what to you is right is not to others. The most drastic change is when you look at yourself in the mirror and see a plump body, which is really thin and scrawny, but you see it as too fat and you deplore being fat.

The other change is that you develop an aversion toward your mother, whereas you have always loved her and the two of you got along very well. If you have sisters, you turn against them too, and all the women in your family. This antagonism is so strong you can't control it and show it in ways that make it very hard for them. You start fights, argue incessantly and refuse to cooperate. You do what you want and disregard their feelings and needs. Life in the family is chaos and you sense that you are at least part of the cause.

The only person you feel is on your side is your father and perhaps your brothers. You complain to them about your mother and sisters, saying horrible things and making up stories that are not true. Your brothers run from you, and your father is put in the difficult position of a buffer between you and your mother, helpless to defend her or change your feelings about her.

Every meal is a nightmare. Although you are hungry, something tells you not to eat. You feel much too fat. Your family keeps urging you to eat and you get mad, make a scene, and leave the table. Everyone is uncomfortable. You upset them with your illogical accusations. Arguments ensue and criticism, and threats fly back and forth. The family is torn apart.

Remember that your illness is limited, and you will be well again, but it will take three to five years. This puts a strain on the family, but it is you who has changed, so please listen, and do what the doctor says. Your parents and siblings love you and it is agony to them to see you starving yourself, so try and understand what they are going through. Although you may not think they make sense, accept it and do as they say. Remember, you are not thinking straight, and most important of all, when you are told to eat - EAT. No one can live without food.

You feel closer to your father than your mother right now, so draw near and confide in him. He is doing the best he can, but cannot take sides between you and your mother.

How do you cope with bulimia? If you have this, it is the same, only the opposite. You are obsessed with food and have to eat. You are furious when told to stop eating so much. But you stuff yourself and then go to the bathroom and throw up. Soon you are hungry and eat again. Does that make sense? So, as with anorexia, listen to your doctor and your parents and do as they tell you, because your thinking is not right. And this you must accept.

III THE PSYCHOSES OF CHILDHOOD

PREAMBLE

The psychoses of childhood are extremely rare and not easy to define. They occur in babyhood and childhood and are recognizable by certain symptoms in each stage of the child's development.

Children can have the major psychoses of chemical origin, the behavior difficulties of electrical origin, but rarely addictive tendencies. They also have physical problems, such as Tay-Sachs disease, hemophilia, and sickle-cell anemia, all of which are hereditary. We see the pre-schizophrenic child, withdrawn, shy, sensitive, unable to get along with his peers, suffering in all his relationships. We see the pre-psychopathic, acting out youngster, aggressive, taking advantage, stealing, stamping on insects, torturing animals and hurting his weaker friends. The pre-alcoholic can be observed in children who eat too much, take too many sweets, drink fluids constantly and greedily grab at oral pleasures.

We can spot in children conditions that will be serious in later life. An example is the adopted child. Why do so many adopted children follow the patterns of their real parents rather than their adoptive parents? Daughters of doing parents are sexually promiscuous at a young age and the boys are acting out and uncaring with no regard for other people or their property. Equally true but less conspicuous is the thinking child in the doing family. Few turn out well regardless of the sincere efforts and loving care of their adoptive parents. Look into the background of their natural parents and this becomes glaringly apparent.

Emotional Deprivation

The most noticeable psychosis in babies was observed years ago in hospitals and institutions such as foundling homes. The babies were listless, lethargic and unresponsive. In the 1940s, studies compared babies with the same background who were confined to institutions and those who were cared for at home. The findings were significant, and indicated that the psychosis was caused by the lack of mothering, loving, holding, comforting. This is emotional deprivation.

To confirm this, another study was done in two institutions, one for wayward girls, who cared for their babies and the other a foundling home in which eight babies shared one nurse. The mothered babies thrived, but at the end of two and a half years, thirty-four of the ninety-one in the foundling home were dead from no discernible cause. These infants had a mournful expression on their little faces resembling depression. All attempts to console them were met with screaming and rejection. They were thin, susceptible to infection and slept poorly, and as time went by they became stuporous and deteriorated rapidly.

To further confirm the need for mothering, the babies in the penal institution were deprived of their mothers for from two to three months. During these periods, the babies developed the symptoms described above. When the mothers returned, the infants recovered completely.

So, if a baby is deprived of mothering and survives to the age of reason, say, three or four years old, the effects are irreversible and last a lifetime. When deprivation takes place after that age, a child can get by with friends and relatives providing the necessary stimuli.

All types of deprivation result in "learned disabilities", meaning that a person may have these feelings but knows they are surmountable by relearning. And deprivation after three or four years of age retards the development of skills and talents and the ability for achievement. There will also be residuals of frustration, despair, anger, depression and feelings of futility, brought on by illness, fatigue or pressures of life.

Years ago things were different. If a baby was without a mother, there were baby nurses and grandmothers to take over. Today, most women are emancipated and care only for their own. There is no longer a maiden aunt or an English nanny to pinch hit. White babies fare the worst, for white people don't help one another in this way. The black community, however, is notable for lending a neighborly hand, especially with children.

Autism
What is autism? Autism was first described in 1943 as a devastating, lifelong disorder from early infancy on for which there is no treatment. An infant can best be diagnosed as autistic by remarks of the mother. "I never could reach my baby. He never smiled." "As soon as she could walk, she ran away from me." "She wasn't cuddly and never wanted me to kiss or hug her." "He always tried to creep off my lap and out of my arms." "She never came to me or appealed for help."

Apparently to feel secure, the autistic child surrounds himself in a tiny world of his own. He sits in a certain spot with a few inanimate objects, which receive his only show of affection. He gives the impression of reigning like an omnipotent potentate, communicating only with the toys he allows inside his little kingdom. To him, the outside world doesn't exist except to furnish him with the necessities of life, to feed him and keep him warm, clean and comfortable. He lets these demands known by gestures.

As a rule, these children are totally mute. If they do speak, it is only to the objects they include in their inner circles. They want to be left alone, this is their defense, and the outside world seems a threat to them.

Parents are frustrated with the autistic child for their inability to get close and reach him. This results first in anger, then guilt. What did we do wrong? But whatever they did or did not do does not cause brain injury which is what autism is, and it has been determined that the wrong kind of parenting does not cause it.

The autistic child appears normal and grows to adulthood, but he doesn't communicate with the world about him, mentally or socially. He has few language and thinking skills and displays behavior such as rocking and banging his head against the wall, or hurting himself in some way. In addition, autistic children, most of whom are boys, are prone to seizures and unusual electrical activity on the surface of the brain.

Do we know why these babies are born this way? Not entirely, but recently the cause has been determined. Researchers found clues to the differences between the autistic and the normal brain through autopsy. But as the incidence of the disease is small and the number of autopsies smaller, there are limited findings. It was found, however, that the cerebellum, the part of the brain involved in muscle coordination and the regulation of incoming sensations, contains fewer neurons than the normal brain, and that growth of sections of the limbic system, which involves memory and emotions, is arrested during development in the womb. Thus, the child is hindered from birth.

Up until now, the controversy of "nature versus nurture" took precedence over the basic cause of autism, but this research definitely points to biological and inborn factors. One may be the over production of serotonin in the brain, another, extensive damage to the cerebellum, a third, marked abnormalities in the limbic system and the cell pathways that connect the limbic areas to the cerebellum. There are strong indications that these changes occur before birth.

We don't know what causes these brain defects in the developing fetus, but it is known that the hippocampus and related parts of the brain are important regulators of memory. This suggests why the autistic have difficulty in social interaction, language and learning.

What is the treatment for these children? Treatment is psychotherapy by a child psychiatrist with the help of psychiatric nurses and child specialists' aides. Dependency on the mother or mother substitute is essential to growth and development which is usually slow and severely limited, and may eventually reach a level of partial function. The cause of autism is physical with brain damage found in most cases. Prognosis is poor, but research is being done to find new techniques to treat autism. No medication is known that alters this illness.

Along with psychotherapy, concerted efforts are begun by all involved with the child to have him stop bizarre and disruptive behavior and to make contact with others and imitate what they do. This is done by firm, but gentle discipline in preschool and kindergarten classes and by the parents at home. It is found that normal children help the autistics, leading them by the hand and showing them how to play. In addition, they receive forty hours a week of training by a specialist in behavior modification, which is continued at home by the parents. There is no medication, but supplements of vitamin B6 are recommended, which appear to enhance the ability of these children to pay attention and learn by observing.

These studies show that if treatment is interrupted after a year or two and the children placed back into their original environment or a mental institution, they revert to their former state. However, if they are sent home, and the parents maintain the training, they continue to make progress. This places a great demand on the parents and not all can stay with it.

A new program brings hope to the autistic child. An intensive behavior modification program from the age of three and continuing for six years has improved ten of nineteen children in one study, to the point that they can fit comfortably into a class of normal children and compete intellectually. Eight others attended special classes for children with language problems. Only two showed no improvement. Such limited experiments point out that our experience is still at a beginning.

These ten children are now teenagers and do not differ from normal children on extensive tests, interpersonal relations, emotional stability, social skills or intelligence. However, this program may not be for all, as there are different patterns of autism and children who are severely compromised fail to benefit.

Not all autistic children have low I.Q.s. They range as in the normal population, but if it is sixty or over, the chances appear to be excellent, and their I.Q.s are greatly improved by the end of the course. Two criteria are essential to success, the intense involvement of the parents and the free association with normal children without the stigma of being "different".

Such reports are encouraging but must be viewed with caution, since diagnosis of autism is difficult, as many cases may represent other physical and emotional disturbances.

Recently, an interesting phenomenon was observed, that there may be a link between high intelligence and autism. Although autism is a rare condition, in about ten percent of cases there is a demonstrable connection.

These few autistics are called autistic savants. The word savant relates to the idiot savant as described in chapter 26.

These are examples. Joseph was autistic, but at four years old, he could draw a map of the world and spell out every country and its capital correctly. Joe is now thirty-one, and when asked what he had for breakfast, he cannot tell you, but he can multiply in seconds 3,458 by 6,927. However, the autistic savant is not as adept as a two year old in social and mundane matters. The brilliance of their achievements is offset by severe mental deficiencies in these areas and some have I.Q.s in the retarded range and can barely speak a word.

A few autistic savants go to college, some graduate cum laude, which earns them skilled jobs. Mark, at four years old, was doing algebra and his combined SAT scores were 1570 out of a total of 1600 by the time he was sixteen. He went to Yale University and graduated cum laude.

Nevertheless, when it comes to social and pedestrian amenities, autistic savants fail miserably. That part of the brain seems to be inert or lacking. From a practical sense, they are incapable of handling the practicalities of every day life or even to be aware of the necessity for them.

David is a good example. He is twenty-seven, lives in his own apartment and has a job as a messenger for an advertising agency. One day his sister visited him and found the floor of the apartment flooded with water from a malfunctioning air conditioner. David was paying no attention to it and when asked why he did not call for help, went into the bathroom and brought out a bath mat. "This always dries," he said. He had lived in ankle deep water for two weeks and could not relate the small problem of the damp bath mat to the large problem of the flood. Thus it is with autistic savants.

CONGENITAL ABNORMALITIES

The word congenital means "born with". Certain abnormal conditions occur in babies while in the uterus, some come from an inherited gene, others from a cause unknown. They are relatively rare and most can be corrected by surgery.

The abnormalities are dislocation of the hip, hernias, pyloric stenosis and various irregularities of the heart. There is the cleft palate, hairlip, club foot, spina bifida and others too uncommon to mention. Today, modern techniques make it possible to diagnose and correct some of these conditions while the baby is still in the uterus, thus giving the fetus the benefit of full term development before birth. Surgery is performed equally successfully on others after birth.

The most common congenital anomaly is Down's syndrome, formerly called "Mongolian idiot" because of a facial resemblance to the Mongol. These babies are usually mentally retarded and are described in chapter 21.

Surgery is of no benefit to these children.
It is important to understand that none of these conditions is the fault of the parents. They are uncontrollable acts of nature.

12 MOVEMENT DISORDERS

Movement disorders of childhood are tics, hyperactivity, attention deficit and one or two extremely rare conditions. They are not common. Only recently has their biochemistry become clear so pharmacological agents can be formulated to treat them.

What is a tic? A tic is a brief, quick, involuntary movement of a small part of the body. It is irresistible and recurs frequently. It serves no purpose and seems out of place. Tics are believed to be habits but this is not so. They are involuntary. The origin of the word "tic" is unknown, but is thought to be onomatopoeic, short and swift.

The face and neck are most frequently affected. A child grimaces, blinks, shrugs the shoulders, or makes quick movements of the head. Another form is wryneck, or torticollis, which is a habit of lifting the chin and rotating the head to one side. Tics are not to be confused with "tic douloureux" or facial neuralgia.

There are three categories of tics, transient, chronic and multiple. Transient tics are also known as "habit spasms" and last from a few weeks to a year. They are self-limiting and require no treatment. The chronic type may start in childhood and continue through life, the multiple type begins as a single or several tics, subsides and disappears in late adolescence. All tics come and go, increase and decrease in severity, especially with anxiety and tension, and disappear during sleep.

What is the cause? Although not confirmed, recent evidence points toward abnormalities in catecholamine metabolism and a disorder of dopamine metabolism, which are considered hereditary. Some tics can be generated by drugs for hyperactivity and by withdrawal symptoms from tranquilizing drugs. These disorders can come from emotional stress, and are made worse if the family tries to force the child to stop. This he is incapable of doing.

What is the treatment? Tics are treated with carefully monitored doses of halperidol, reinforced in selected cases by clonidine.

What is attention deficit and hyperactivity? Attention deficit means difficulty in adapting to the structured setting of the classroom. It is usually accompanied by hyperactivity, but may not be. The attention of these children is short, and they have learning disabilities which are intensified because they are easily distracted.

It is estimated that five to ten percent of school children have attention deficit disorders. Males are predominantly affected. Symptoms range from impulsiveness, excessive physical activity and difficulty in perception, conceptualization, memory and language. Teachers have trouble keeping these children from disrupting the class and are forced to turn to the physician. It is necessary to medicate about three percent in order to control their behavior. This relieves the classroom situation, but fails to uncover the cause of the child's problem.

What is the cause? Heredity (genetics) is the most common cause of attention deficit disorders, with or without hyperactivity. Genetic conditions that contribute can be epilepsy, depression, difficulties in adjusting to school situations, reaction to medication or allergy to food. There are organic causes, like injury to the head from accident or infection, an undiagnosed brain concussion or a major psychiatric illness.

Other factors are environmental. A life event that creates anxiety, such as the loss of a parent or a serious illness or the wrong educational setting. Attention deficit can result when more is demanded of a child than he is capable of. A high level of lead in the blood can also cause this disorder. To make a diagnosis, the child is closely observed. A careful history is taken and analyzed, and every possible cause ruled out by thorough physical, and psychiatric examination.

What is the treatment? Hyperactivity is treated by drugs, and the appropriate dose is determined by the cause. Anxiety states require mild tranquilizers, depression is treated with anti-depressive drugs and psychiatric disorders with anti-psychotic medications. Research shows that some hyperactive children have low amounts of dopamine, which can be increased by drug therapy.

Learning remains below normal level, so a one-to-one teaching program with behavior modification by a qualified therapist is recommended.

HOW DO YOU COPE WITH THE DISORDERS OF CHILDHOOD?

If you have a child in need of psychiatric help, don't despair and don't let him know that you are upset. Treatment is available, it is effective and your child can be helped. Whether it is a tic or a movement disorder, many people believe these conditions are serious, and family and friends accentuate this fear, sometimes suggesting that they could be fatal. This is not so. Listen to your doctor.

These conditions are not permanent and not disabling, and children can live their lives in a normal and healthy fashion. It is important, however, to have medical and neurological advice and to consult the pediatrician or family doctor, who will suggest specialists to see. A psychiatrist and a neurologist will rule out organic disease and make the correct diagnosis. Once this is done, the child can be treated successfully.

IV UNDERSTANDING DISTURBED BEHAVIOR
Personality Disorders From Electrical Problems of Doing

PREAMBLE

In Chapter 2, we discussed thinking and doing. We have just told you about the various psychoses that arise as a mental illness in the more thinking, in medical terms, the more schizoid individual. The following section is devoted to personality disorders in the more doing person, the more psychopathic.

These are the people who are impulsive and without conscience. As you have already learned, their behavior is of an electrical origin, whereas, a psychosis is due to a chemical imbalance in the brain, and can be treated with chemical substances, but as of today, research has found no treatment for personality disorders, although every known therapy has been tried.

Much research has been done and is still ongoing to map the electrical circuitry in the brain in an attempt to analyze it and find means to alter or control electrical impulses. Presently, the only method we have is to apprehend the wrong doer and incarcerate him, but unfortunately, the criminal justice system is not geared to keep criminals in jail and the rate of recidivism is extremely high. In a majority of instances, when a subject has served his term and is released on parole, he breaks the parole and commits another crime. The following three chapters explain these disorders.

13 THE BORDERLINE

Three hundred years ago, Thomas Sydenham, an English physician, described borderline patients, then called "hystericks", with: "All is caprice, they love without measure whom they will soon hate without reason". This phrase still applies to these unusual cases.

The borderline patient is neither fish nor fowl. He is sufficiently normal to live and function in the world of reality, but at the same time borders on the psychotic, the neurotic, and the psychopathic, all three, never showing only one condition. Hence, the "borderline".

His hallmark is an abnormal twist in personality and the inability to experience feeling. He doesn't care about people, nor is he sympathetic to their misfortunes. He has both doing and thinking elements so interwoven that his emotions are totally subjective, thinking only of himself. He has impulsive swings from love to hate, acts out illogically and impulsively and is often addicted to alcohol and/or drugs.

The borderline's relationship with others is tempestuous and unstable. He is obsessed with fear of being abandoned, so he clings to one person and cannot be left alone. He is capable of intense anger and lack of control, and may commit acts of self-mutilation. He is the doctor's most difficult patient and often pesters him with late night calls, or appears at his home or office at all hours.

Depending on the symptoms, borderline cases are difficult to identify. They are classified as paranoid, schizoid, depressive, hypochondriac, psychotic, anti-social or mixed. They are manipulative and vacillate from one symptom to another, such as from affection to intense anger, from passivity to violence. Unlike the psychotic, they are contriving and use illness to get what they want.

Borderline patients are exceptionally shrewd at achieving their goals, but they show extreme instability in interpersonal relations, mood and self-image, which usually defeats their purpose. They are generally antisocial and cannot relate to others because of distortion in their thinking. They may be highly intelligent but incapable of concentrating for long, which makes them erratic in their careers. Borderline patients are likely to depend upon their families or drift into unemployment, alcoholism or crime, depending on their personalities. Their judgment is poor, and they have little common sense. Sexual function is distorted into too much or too little, and they suffer from impotence and frigidity or excessive sexual drive with promiscuity.

It is not easy to diagnose the borderline. This is made primarily on the past and present behavior of the patient and shows up early, usually in the teens, and continues through life, mellowing somewhat after the age of fifty. Up to that time, there is a persistent pattern of substance abuse, emotional turmoil and aberrant personal, sexual and financial behavior. The borderline shows a poor choice of friends and activities, extremely poor judgment in all major areas and a great fixation on himself and his needs. However, these are not consistent, one week he wants to spend money, the next to save, then he will buy an extravagant item he doesn't need and the next week start saving again. Nothing he does makes sense and his behavior is consistent only in its inconsistency.

He makes demands on other people and misuses their feelings, and whatever they do for him is not enough. He has to have more. He has an inexhaustible well of demands, an insatiable drive that is never satisfied.

The borderline illness can be called a mongrel psychiatric condition. There are common characteristics, one of which is absolute dichotomy of thinking. Everything is black or white, good or bad. He won't tolerate a neutral statement or a double meaning. He is unpredictable, especially in matters of money, sex, overeating, gambling and physical self-harm. His relationships are fleeting and characterized by rapid shifts in attitude from tolerance to exploitation. Mood swings and suicide threats are almost always present, but the borderline rarely carries out these threats successfully.

This type of patient has a low self-esteem and his opinion of himself is confused about identity, goals and values. His mood is unstable, shifting from high to low at a moment's notice and switches from a regular mood to depression, marked with excessive irritability and anxiety of hours or days duration. He hates to be alone and goes to extremes to avoid solitude, which makes him deeply depressed or intensely bored.

Here is a typical case. Penelope is well-to-do and commands the support services of twelve people, none of whom know that there is another professional treating her. She engages a social worker or two, several nurses and consultants, three psychologists and two psychiatrists to whom she goes alternately for therapy. She spends most of her time shuttling from one to the other and never tells any of them that she is being treated by someone else or how. She operates skillfully, manipulating to get attention from them and anyone who will listen, using different symptoms and demanding respect and adulation from all. In addition, she abuses pills and alcohol, and performs irresponsibly with her friends and family. She also makes repeated suicide threats.

The borderline's thinking is split. He places a rigid dividing line between positive and negative thinking, and between feelings and opinions. A normal person can sit on the fence and see both sides of an issue, but the borderline cannot see both sides at the same time, although he can switch back and forth when he chooses.

Occasionally these patients show symptoms of neuroses and experience psychotic episodes, and have delusions like those of the schizophrenic, which can lead to an incorrect diagnosis. However, they quickly regain equilibrium, whereas the psychotic does not.

From a psychiatric standpoint, these people are neither neurotic nor psychotic, but poised between psychosis and psychopathy, that is, between a thinking and a behavioral disorder. They have fluctuations in mood that frighten them, and they learn to use symptoms rather than treatment to help themselves. They suffer from an unusual kind of hysterical mood disturbance and from weak self-esteem and self-image.

The cause of borderline illness is unknown, but heredity is largely involved and the environment may also be an influence. The theory is that these cases border on effective disorders and to some extent schizophrenia, both of which have scientifically proven genetic sources. Hence the name "borderline". The same studies on borderlines suggest genetic factors in some, but have been inconclusive to date. One ongoing study reveals that in many cases the borderline might be brain damaged from repeated physical abuse as children by parents and/or serious accidents requiring emergency room visits.

Can it be treated? This disorder is chronic throughout life. Its main characteristic is that the patient remains substantially the same, neither better nor worse. He is stable in his instability. Appropriate drugs have been tried on the variety of symptoms, but most have proven ineffective. Some work on certain symptoms, but so far there are no double blind studies, because the standard diagnostic criteria are new. Drugs of limited benefit include minor tranquilizers for anxiety, anti-psychotic drugs for psychotic attacks, anti-depressants in selected cases for insomnia and anxiety. None of these medications is prescribed on a long term basis, but only as symptoms arise.

Psychotherapy is debatable because so far, the illness has been found to be incurable. Long term, intensive psychotherapy with the goal of changing the patient's self-concept offers hope of improvement, but there is no guarantee. Supportive psychotherapy with drugs is effective for severe transient episodes, but not in the long run, however, it has its place for certain patients, as does group therapy. In short, we need new methods and new drugs, but most of all more research into the cause of the disease, with an eye to preventing and eliminating it.

The borderline personality is uncommon, comprising one to five percent of the psychiatric patient population, but it commands the attention of psychiatrists because of its complexity, its dramatic and conspicuous features, the severity of its symptoms and the distressing effect on the family. Technically, borderlines are intermittently verging on illness, most often toward the psychopathic or doing side. Occasionally they verge on psychoses, and when they are stable in their instability, they are described as neurotic.

How do you cope with a borderline? The borderline is a difficult person to live with. Whether you are a parent, spouse, child or sibling, first of all, learn to protect yourself. Loving parents are victimized the most and they particularly, need help, for they cannot escape. Brothers and sisters seem able to take care of themselves, so parents really bear the brunt, as they are responsible and involved the most deeply in the process of handling the patient, who looks for the kindly soul, latches on and uses him or her to the extreme. These victims are the parents, the kindly physician, the caring cleric or spiritual advisor, the responsible official in organizations like the Red Cross and the Welfare Department.

The wife or husband of a borderline is truly miserable and the marriage usually ends in divorce, for it is more than a spouse can tolerate. If, however, he or she is determined to keep the marriage together for personal or religious reasons, first, he must protect himself by becoming insulated, and go on with his life apart from his spouse. This is the best advice to give the mate of a borderline, because they create constant conflict, pressure and demands, repeated suicide threats and high expenses of medical, psychiatric and surgical treatment. There are vehement ups and downs of mood, wide variations of behavior, abuse of alcohol and drugs and continuous manipulation.

Second, he must understand that the patient is going to change rapidly. There will be changes in actions and relationships, so it is impossible to find a way to handle him. Moreover, his actions are unpredictable. There are many suicide attempts, sometimes serious when in a low mood, but borderlines rarely kill themselves. In high mood, he is especially dangerous, not in a suicidal way, but he may spend money he doesn't have, get married, rush into an impossible situation or try to do something foolish and attention getting.

The borderline has mood swings, high and low, much like the manic depressive. In fact, a new school of thought believes some borderlines are manic depressive with a psychopathic personality and should be treated as such. They are addictive, immoral, with poor interpersonal relationships and poor judgment. They never keep friends but are tireless in seeking attention. Attempts at suicide are one way, and the addictive tendency is another. They try anything and everything, alcohol, drugs, glue sniffing, any kind of pills they can get their hands on and in large amounts. All this is mixed in with mood swings.

Another way the borderline gets attention is to break the law. He delights in being picked up by the police or put in jail for not paying a bill or pilfering some useless article. He presents a challenging attitude and loves to buck society and its mores and laws. He can be very destructive and bring out the worst in others, so the family has to protect itself against legal onslaughts and public retribution.

All of a sudden, your borderline makes a complete turnabout and is sweet, accommodating, supportive and wonderful. During these interludes, his real personality emerges, but unfortunately, this stage is short. There is a chameleon-like quality that seems normal, then suddenly, he is just impossible. This is very baffling and extremely disturbing. Who can understand the borderline? Even psychiatrists can't.

14 THE PSYCHOPATH

What is a psychopath? The psychopath is the person you hear about in the press who has committed an atrocious crime. This man has an antisocial personality disorder and is born with an extreme doing personality. Very early in life the traits that earmark him were apparent.

Recently, there was developed a list of criteria for the diagnosis of antisocial personality disorder, namely, the psychopath. The list consists of overt behavior beginning in childhood of lying, cheating, truancy, fighting, cruelty to animals, vandalism and so forth. And in adulthood, it consists of reckless driving, promiscuity, frequent changes of jobs, illegal activity and irresponsibility of financial and family obligations. In addition, the compilers of the list include twenty items. They are glibness and superficial charm, arrogance, lack of realistic long term goals, inflated self-esteem, manipulativeness, lack of remorse, callousness, impulsiveness, irresponsibility and shallowness of feelings. Two sets of characteristics that are highly interrelated are impulsiveness and instability, callousness and egocentricity and a limited capacity for anxiety.

When a person has a high score on the check list, it is not correlated with intelligence or neurological abnormalities. However, it is associated with drug abuse, alcoholism and the absence of anxiety, depression and other psychiatric symptoms apart from personality disorders.

Called the Psychopathy Checklist, it predicts fairly accurately whether a person will remain in therapy, work at it, and show improvement. High scorers usually don't. It also shows that a high score on the checklist is a good indicator of future violent crime and more accurate than past crimes committed. It also can predict the type of crime the person will commit in the future, and that it will be more cold-blooded, less impulsive and more likely to be directed at anyone — men, women, intimates, strangers.

Let's go back to the young psychopath. As a boy, he is ruthless in his treatment of parents, siblings and friends. He is destructive, cruel to animals, and oblivious to learning right from wrong. As a teenager, he vandalizes the neighborhood and may become a runaway, a drug addict, or both, supporting his habit by theft. In adulthood, the scope of his activities may extend to arson, rape, assault or even murder. He is totally antisocial and has no regard for the consequences of his acts either to himself or to others, and takes pride in the suffering he inflicts. His ruthless nature overrides all feeling of compassion and he has no understanding of the anguish and grief he causes. In fantasy and in real life, the psychopath concentrates on gaining his pleasure, regardless of the harm he may inflict on others. He is cruel and demanding and experiences no remorse after hurting someone, stealing or destroying property. This is the psychopathic personality.

The psychopath inwardly gloats over his accomplishments and enjoys causing havoc and pain, which he observes from the sidelines with amused detachment. He delights in setting fires and mingling with the crowd watching the blaze, the havoc he has created and the property he has destroyed. Sympathy is unknown to him, as he is totally self-oriented. He has no conscience and is incapable of emotional attachment, but may take part in family and social life in an aggressive and selfish manner.

His doing personality is the conscienceless type and drugs and alcohol compound his viciousness by releasing all vestiges of inhibition. While buying a pack of cigarettes, he delights in scheming how he can rob the store or abduct a woman and rape her. He may act impulsively even when there is a risk of being caught, for he takes pride in eluding the police and boasts about it to his friends.

These men are unable to support themselves and become burdens on their families and society. They seem unable to hold a job for more than a few months and are irresponsible employees, come to work late, take days off without reason and don't care about job performance. They are prone to initiate fights and may use a weapon to inflict injury on their opponents.

Psychopaths are extreme doers, and some have strong paranoid trends. The affluent, intelligent, successful person who is a psychopath will be as free of moral restraint as an ignorant man who is a failure.

Some criminals marry but never remain in a relationship for long, a year is the longest. They are extremely promiscuous from teenage on and liable to abuse their wives and children. They are known to force sexual intercourse when the urge arises. The victim could be a child, a relative, an acquaintance or a stranger he covets. They fail to pay bills they have incurred or continue child support by moving to another location, and are inclined to move from place to place and never establish a fixed address. They are often incarcerated in penal institutions for their misdeeds. The death rate is high and many die by violent means.

Can we predict what will happen to the young psychopath in later life? An interesting study in this regard was recently completed of fourteen boys sentenced to death for crimes committed before the age of eighteen. These young men on death row came from four states. Six were black, seven white and one Hispanic. They committed their crimes at an average age of sixteen, and during their trials they never revealed their backgrounds of abuse and psychiatric disorders, but preferred to be considered as criminals rather than abused, psychiatrically impaired or of low intelligence.

The study consisted of psychiatric interviews and other thorough examinations and tests and histories of psychiatric disorder and family violence. Revealed was physical abuse of eight boys with injuries requiring hospitalization, nine had serious neurological deficiencies, one with grand mal epilepsy, seven were psychotic, four had severe mood disorders and several had paranoid tendencies. Only two had I.Q.s above ninety and the abstract reasoning of ten was severely impaired. Twelve of the boys had been brutally abused and five sodomized by relatives. It turned out that all of their parents had a high rate of alcoholism, drug abuse and psychiatric hospitalization.

The career criminal is a psychopath. He rationalizes continually, and excuses his behavior by setting his own rules of conduct. Sticking to them results in a fixation of the pathways between thought and action, so that eventually he excuses everything he does. Most of the time he appears to be like anyone else, and others are fooled into thinking he is. But his normal behavior is broken by episodes of spontaneous acting out for immediate self-satisfaction and gain. The criminal mind is clever and rational, but cannot be controlled once the mold is set in the brain's circuitry.

A criminal is usually of medium to high intelligence and energy, and plans his strategy so he won't be caught. If he is caught, he boasts that he helped the police find him. He really believes his excuses and lies and can't understand why the authorities don't. If he is urged to confess, he will go to any length to deny his crime. If he does confess, you cannot be sure he is telling the truth. Sometimes he will confess to a crime he hasn't committed and later retract his statement, claiming he was coerced.

While we are all manipulators to some extent, the psychopath manipulates with malicious intent. He steals what he wants to fill his needs, both emotional and physical through lies, bullying, blackmail, and other tactics. Unlike the thinker, he never becomes involved in ideas or activities beyond himself. The less intelligent criminal relies primarily on brute force to gain his objectives.

Some career criminals form groups under a set of principles to justify what they do. They consider themselves a separate society with different but equally valid standards. They are the masters of organized crime, and have formed this subculture, whose members make their own laws and adhere to them under their own death penalty. Their emphasis is on loyalty to the group.

Disgruntled ethnic groups adopt the same idea. They assert that they are justified in seizing their share of the world's goods by force. Individuals also presume that they are entitled to more than they have, even though what they have is plenty. The poor have no monopoly on grasping attitudes. Some of the world's wealthiest continue to amass wealth and everything they can by any available means. Rich or poor, all such people have psychopathic traits colored by high paranoia and closely linked to the male sex drive. Few are women.

On the other hand, all psychopaths are not criminals. Woven throughout the fabric of society are threads of this kind of person in all walks of life. They are not criminals, but have a criminal's inborn characteristics. They feel they don't have enough of what the world offers, even though they have more than they need. These grasping people have a materialistic complex that cannot be satisfied. It comes from within and is linked to a feeling that they are being mistreated.

There is always some thinking mixed into the psychopathic personality from his thinking side. Although his life is directed by his doing, he is capable of warm, loving relationships with those close to him and can be friendly and outgoing. However, this emerges only when he is relaxed, and is seldom evident to the outside world. But when he is under pressure, watch out! Then his true colors show. Then, he thinks only of himself and what he wants.

Some scientists maintain that psychological factors are the source of this type of behavior. Emotional deprivation, the lack of love in early life is the commonly accepted cause. The second is depression, which makes this kind of man feel that what he has is not enough. Third, excessive paranoia creates dissatisfaction, fed by the affluence in which most of the western world indulges. Everyone has a car and is taking a trip or buying a house or some expensive luxury item not really necessary. We are exposed through television and bombarded by the media with things to buy and to do and the stores are brimming with tempting goodies. However, these theories are invalid when environmental and hereditary factors are traced. Inborn psychopathy is the predominate cause.

The most dangerous criminal is also extremely paranoid as well as psychopathic. He is the paranoid psychopath. Adolf Hitler, Stalin, Mussolini, and in later years, Ceausescu and Saddam Hussein are examples. There are several others we know of making trouble today, each in his way out to destroy the good in the world. There is nothing you and I can do but let them run their courses, and they will. Fortunately, they are rare.

The paranoid psychopath has high energy, high intelligence and is an extreme doer. He is outgoing, power wielding and ruthless, and can influence an entire nation, as did those mentioned above. They are the evil movers and shakers of the world, motivated by their compulsive drive to control, whether it be political, or religious, or financial.

They are destructive, dominant people who attract great followings and sway strong personalities. Not all are master criminals or dictators. Many go into legitimate businesses and run them in an illegal way, unaccountable to society and the law. They are smart and capable and powerful, selfish to the extreme and care nothing about anyone except themselves. They are clever and manipulative, and lack all feeling. There are several examples in this country today, as aptly recounted by the press. They are called "white collar" criminals.

Are paranoid thinkers and doers different? There is a wide gap between the paranoid psychopathic doer and paranoid psychotic thinker. Both are hostile, suspicious and controlling. But the thinker, even when he seems aggressive, is a coward at heart, whereas the doer reacts to fear with aggression, and is a danger to the community. The thinker has moral restraints. The doer has none and is a consummate liar as well.

The sexual and the aggressive areas are located close together in the brain, and can act simultaneously, and often do. Excessive paranoia can make both men and women dangerous. Paranoids interpret everything as they wish, and always to their advantage. They are not good listeners, but love to hear themselves talk. They often emerge from the masses and rise to a dangerous height of influence and control. They can be powerful and charismatic and spellbind millions.

There are also psychopathic women who operate in a different way, covert, hidden. They are not as aggressive nor as violent as men. Women are also more verbal in their psychopathy, which comes out as sneakiness, treacherousness, manipulativeness, deceptiveness, back biting, undercutting and disparagement.

These women also show immoral behavior, illegal, antisocial, lying, cheating, stealing and so forth. They have the extreme doing type of personality that is insensitive to the needs and feelings of others, and only interested in their own. They know what they do is wrong, but rationalize: "Well, I did it, but I don't care. So what?"

There are classic cases of manipulative psychopathy in women in popular soap operas currently on television. Another example was in the newspaper. A woman's husband was reported to have drowned in a boating accident, and when his body was not recovered, an investigation ensued. This was the woman's fifth husband, all of whom had died. Eventually, his body was found buried in the lawn of her house, and that of the fourth husband next to him. She is a paranoid psychopath.

We all have paranoid qualities in common with lower animals, but extreme paranoia is peculiar to man, whose evolution took a different turn in this respect. Both men and animals fight, but an animal fights only for food or to protect its young. Man fights for ideologies and goes on killing after he has won, to impose his ideas upon others. Excessive paranoia takes distorted forms under the influence of intelligence and energy, changing it into an unpredictable source of anger and hostility. Thus, man is called the most dangerous animal. Fortunately, only a tiny percentage of paranoid people are psychopathic killers.

Every day we hear of senseless crimes and wonder what is behind them. A man runs amok in a tavern and shoots eight people, another beats his eighteen months old son to death. Although uncommon, it is not unknown for a woman to commit a crime of violence, usually against a child or her husband. In terms of lovers, it is usually a "crime of passion".

There is violence in everyday life. It is natural for boys to fight, and most fight in fun, but a bully fights to control, and there is no limit when he gets the upper hand. He tells his victim to say "uncle", and pummels him until he does. This is a tendency to hurt in order to get one's way, and is the primitive instinct which can turn into savagery. Control is a heady stimulant.

This destructive drive is also seen in vandalism. When adolescents break into a school and smash things, they do a thorough job, broken windows and fixtures, splashed paint, flooding from fire hoses. The thinking element disappears, and what is left is a thoughtless, primitive force running out of control. The same instinct takes over when a mugger holds a knife to a victim's throat, takes his wallet and jewelry, then kills him. The media offer repeated examples of the robber who holds up a store, empties the cash register, wrecks the place, then kills the owner and anyone who happens to be there. The killings serve no purpose.

Non-violent conflict is governed by the same fierce drive to win. The contest is equally intense and primitive between two intellectuals in a chess tournament. This is also true for sports like tennis, golf and baseball, and games such as football, which are carefully monitored to prevent injuries. What matters is to win and competition is the drive to prevail.

Sudden violence is poorly understood. We know that there are anger and rage centers in the brain. One of these is located near the sexual center, which leads us to believe that violence is tied to sexual aggression by diverting sexual arousal into the rage center in those persons who kill without reason. Their victims are usually women, and they repeat these crimes until they are caught. The Son of Sam case is an example. Between fits of rage, which come on abruptly and leave just as suddenly, these psychopaths appear perfectly normal, making it difficult to suspect, much less apprehend them.

Meaningless killings are the result of imbalance in the control centers of the brain which can come from several sources, like a faulty gene, brain toxicity or injury. They are, however, rare. Brain toxicity is the most common form, often related to drugs. A concussion, skull fracture, or brain tumor can also change behavior. An example is Alonzo, the Texan, an Eagle Scout and an admirable young man, who, without provocation shot at pedestrians from a tower, wounding several and killing two. The police had to shoot him and autopsy revealed a brain tumor.

Normal inhibitions disappear under the influence of alcohol or drugs, the rage center is touched off and the restraint mechanism is temporarily obliterated. In these circumstances, it appears to be a physical rather than an emotional defect that triggers violence, but we cannot be sure. The role of the rage centers is still not well understood.

Where do criminals come from? Observation, experience and research indicate that criminals are born, not made, and they don't change. Once a criminal, always a criminal. Exceptions are those suffering from brain injury or toxicity as explained above. In scientific parlance, these individuals have an antisocial personality disorder.

Many people believe that the environment of childhood is responsible, and a ghetto upbringing is bound to produce a thief or worse. We believe this concept to be false. All men brought up in ghettos are not criminals, they show up as well in well-to-do upper class families offering every advantage. However, although there are exceptions, the majority come from the lower class population and grew up with one parent or in a foster home. The absence of consistent, parental discipline contributed largely to the negative environmental effect. Moreover, it is apparent through studies that most psychopaths had fathers who were also psychopaths. This is confirmed in adoption studies that parents with psychopathic tendencies increase the risk of their children having it also. This was found to be true of both adopted and natural children.

Therefore it is hypothesized that the cause of this behavior is partly due to the environment and partly to heredity.

Several years ago, studies from St Elizabeth's Hospital in Washington, D.C. give unquestionable evidence that criminal tendencies are inborn, showing up in young children and continuing throughout life, and are untreatable by any known therapy. Therefore they are not reversible.

Doctors Yochelson and Samenow devised a technique similar to that of Alcoholics Anonymous, whereby criminals participated in group therapy daily for many hours without a break. The rules were strict, and all were required to conform. The slightest deviation was forbidden, and if a member so much as took a cigarette or a pencil from a companion, he was punished as though he had committed a crime.

Yochelson and Samenow worked for twenty years with two hundred and fifty-five subjects, of whom only thirty completed a program of five hundred hours of group and individual therapy. Of those thirty, ten were regarded as having overcome their criminal tendencies. To our knowledge, there has been no follow up.

However, the high cost of professional time and the meager results of these studies fail to justify this technique for the rehabilitation of hard core criminals. Modern society demands scientific proof and this research is the first. It produced three huge volumes of data confirming conclusively that the psychopathic personality is born and not created by the environment. It also proves that rehabilitation by psychotherapy or with drugs has no effect on the criminal mind.

Does juvenile delinquency lead to adult crime? Not necessarily, A recent study by the Department of Justice shows that most teenagers outgrow delinquent behavior in the process of maturing. About twelve percent go on to adult crime. Conversely, some of the most serious criminals have no police record before the age of eighteen. Thus, juvenile delinquency is a poor predictor of a criminal career.

This same study also concludes that children from broken homes are not more likely to get into trouble. The structure of the family has little relationship to juvenile delinquency, in which the more doing personality is also apt to be involved. Moreover, it shows that those who don't belong to the twelve percent can be changed through discipline and the process of growing up. This confirms that not all extreme doers are bad.

What goes wrong in the criminal mind? Just what and why is a mystery. However, it is certain that there is a malfunction or dysfunction in the communication system between the brain cells, probably electrical in nature. It could be several genes working together, but it is not a chemical imbalance, as in the psychoses, for research has proven that drugs that help the psychoses have no effect on the psychopathies. However, it emanates from the thinking section of the brain, because unnatural thoughts come to us all, but we don't carry them out as the psychopath does. The restraint mechanism prevents this. The restraint mechanism is a part of conscience.

Why is it unsafe to walk the streets? In some cities, it is actually dangerous. And why are there so many robberies, rapes and murders? Almost every day, there is an account on television or radio of another child or pedestrian being shot dead or wounded. The individual for whom the shot was intended often gets away scott free. The police do their best but they are restricted by laws, and several of them have been killed while on duty.

Last year, a couple and their two sons were assaulted at gun point by a gang in the New York City subway. They had come from Switzerland to watch the tennis matches. One son was shot dead when he tried to help his mother whose jewelry was being yanked from her body. In September, there were two programs on television of interviews in prison of two killers. When asked the following questions, both men replied "No" or "Not really." "Are you sorry for what you have done?" "Have you remorse for having killed a child caught in the cross fire?" "Do you pity the mother of this little boy you have killed?" They both admitted to having murdered several and one said he had killed thirty people, responding in a voice tainted with bravado and pride. And it goes on and on.

Drugs are a large part of the problem as the cocaine and heroin addicts kill to get money to support their habits, and gangs roam the streets looking for a fight, as drug lords battle to keep their territory. The government is trying hard to stem the flow of drugs into the country, although the amount has been reduced, the drug lords manage to sneak in substantial amounts illegally by devious means.

Another cause is the prevailing attitude that anything goes. Young people are blind to discipline and punishment. Parents have a role in teaching inhibitions, but that role has been abandoned. Parents are afraid to discipline their children. The schools are legally restricted from disciplining students. The courts, police, judges and governmental agencies are attacked if they use repression or force. Society no longer insists on imposing restrictions, nor does it appear to believe in them. As a result, borderline psychopaths have less restraint and move closer to slipping over the dam. Stronger laws could prevent this in many instances. However, laws do not stop the true psychopath, they only decrease the downfall of marginal cases. The thinking half of society doesn't need laws. Thinkers are self-regulating.

CRIME IS DIVERSE AND TAKES MANY PATHS.
HERE ARE EXAMPLES.

Recently in a large city, there occurred a rash of burglaries of airline ticket offices. An armed bandit forces everyone to the floor, takes the cash and flees. This happened about every five days, until the police staked out the likely targets. The thief was caught, identified in a line up and jailed. The robberies ceased.

After forty years of marriage, a man divorces his wife for another woman. He was always a model of integrity, but recently took a family heirloom from his wife's room and denied it. He has been deceiving her throughout their marriage, while deluding himself that he is a man of high principles. He has destroyed his marriage, driven his wife to the brink of suicide, and disrupted two families. His only regret is that he is being found out.

A man is booked on a minor charge. While waiting in the police station, the officer leaves the room for a moment and he steals a wallet from the pocket of a coat hanging on a chair, removes the credit cards and tosses it under the desk. The officer discovered his loss only after the man paid the fine and left. The psychopath is very clever.

A man accused of shooting a woman is referred to a psychiatrist by the courts. He is an impulsive, brilliant, creative and hard working man, with high principles and rigid ideas. He comes from an intelligent, capable family with no history of violence. He has no psychological aberrations and no brain injury, nor does he drink or take drugs. He is more of a doer than a thinker, which shows up in an occasional gambling bout or when he disappears for three or four days.

After several years of intensive psychotherapy the psychiatrist found no sign of mental disorder. However, the investigation proved by tracing his car that he had killed the woman just before going for treatment, so his crime appears to stem from a combination of the aggressive and the sexual drives which overwhelmed him and forced him to kill. This case demonstrates how little psychiatrists can do for the psychopath.

A policewoman is about to arrest a suspect for grabbing a gold chain from a girl's neck. The robber overwhelms the officer, kills her with her service revolver and flees. He later was picked up at his home where the revolver and chain were found.

How do you deal with a psychopath? If you are involved with a psychopathic personality, you may not be aware of it at first, for there is a wide variety. They range from the totally immoral and the career criminal, to the slightly immoral, the pilferer, the dishonest and deceiving. Then there is the asocial, the egocentric, non-caring, sadistic, to the mild forms of selfishness and self-centeredness.

The psychopathic personality is warm, friendly and outgoing. He makes you feel comfortable, such as putting an arm around you when he scarcely knows you. He likes people and seems to understand them. When he is young, he fears nothing and will charge into any situation, no matter how threatening it is. He tries everything, takes any kind of drink or pill for kicks, and experiments physically and emotionally with almost anything.

He can be a con man, a cheat, a thief, a criminal. He can be an alcoholic, a drug abuser, a sexual offender, a child molester. But he is lovable in a way that makes him appealing. Although, his has a doing personality coupled with poor judgment, his outgoingness gets to you. He is impulsive and quick to act, therefore, families, relatives and friends should understand this and love him only for his lovable side.

But be aware of what a psychopathic personality is capable of, and remember that you cannot trust him. Beware if he asks you to lend him money, don't sign his notes, don't get involved with him in business. And don't trust him sexually. Watch out for what he says, because what he says is not what he does. Believe what he does and forget what he says. He is a fast talker and will promise anything and deliver nothing.

The greatest danger is to put your trust in someone like this, whether it be someone you marry, your child, a relative, friend or business associate. If you do, you will wind up behind the eight ball. Ultimately, he will betray you, and prove to be worthless, indiscriminate and irresponsible.

The psychopathic personality is in every community. Look around you and be aware. If you find one in your family, deal with him carefully and from a distance. These people do not change, nor can they be treated, for their problems are electrical, not chemical. Furthermore, they reject psychiatry and any notion that they are different. A relationship with one is bound to end up in trouble. They are selfish to the extreme and care nothing about anyone except themselves. They are not interested in being helped or in changing their behavior, they see nothing wrong with it. They just keep charging ahead, inflamed with the need to control.

AN UNUSUAL PSYCHOPATH
Munchausen's syndrome is an unusual form of psychopathic behavior. It is a rare but serious disease and difficult to identify. The name comes from Baron Karl von Munchausen, a German cavalry officer in the 1700s, who charmed his guests with yarns about his travels and campaigns. A book was written about them which became popular and remains a classic. In 1951, a British physician used the Baron's name to describe exaggerated fabrications of illness. Hence, Munchausen's syndrome, a fascinating condition.

What does this person do? A Munchausen psychopath's primary objective is for attention. He gets it by feigning serious illness. He has a powerful personality and tremendous talent, but it is used only to further this need. He is intelligent, and has learned medical terminology and the symptoms of the illness he wants to simulate, and how to present them so a doctor believes him. The Munchausen patient is the champion manipulator of the world.

Although his objective is for attention, he is not like the hypochondriac who truly believes he is ill, or the malingerer who fakes illness for a purpose or to sue the physician for malpractice. The Munchausen patient shows up in the hospital emergency room and says: "I have a bleb on my lung and it ruptures and bleeds from time to time, So don't you think I should have a tracheotomy?" Or he may cut his mouth and spit up blood and claim he has internal bleeding. The doctor orders elaborate tests and the patient is hospitalized, which is just what he wants. Then he is put through elaborate tests and will tolerate any procedure willingly to accomplish what he is after. All he wants is attention.

A Munchausen patient is male as a rule, and the age ranges from teenage to elderly. He has personal charm and displays a sense of drama. He bursts into a hospital emergency room and announces that he is dying, insisting that he needs this test or that drug. Examination reveals many surgical scars on his body, and he talks incessantly about the illness he has had. Remarkable for one who is dying!

When the doctor finds nothing wrong, and suggests he see a psychiatrist, he becomes belligerent and hostile. Then he disappears without paying his bill.

The pattern is to go from hospital to doctor to hospital, taking valuable time needed for ill patients and adding to the cost of medical care for all.

Is there a treatment? The Munchausen patient is the most difficult psychiatric patient to diagnose and treat. The number is not known, but cases are reported throughout Europe and the United States. Although these patients need and deserve care, there is no known cause or cure. Treatment with drugs is not effective, everything has been tried in vain. Psychotherapy helps only rarely as these patients live by illness that they are able to feign most adroitly, fooling doctors and staff repeatedly. Thus, they do not want a cure and avoid treatment that might assist them.

15 THE SEX OFFENDER

There are two kinds of sex offenders, those who are violent, aggressive, sadistic and extremely dangerous, and those who are merely an annoyance.

The first group is small and largely males, but they are seldom caught, for they cover up their crimes, so it is not easy to trap them. Furthermore, their victims often are loath to face a trial and help the prosecution. A minority are murderers, but all have amoral personalities.

These criminals have extreme doing — in medical terms, psychopathic — personalities and have slipped over the dam. When they have high paranoia in addition, they are more aggressive and hostile, but very clever at evading detection, so it is hard for the police to ferret them out. Once they are discovered, other crimes they have committed are often revealed.

The histories of these criminals show persistent repeat offenses, and they should be removed from society. Prison does not stop them, because they get sexual pleasure from their perversions, which are easily satisfied in jail. Few seek professional help, except when forced to, and psychiatry's experience shows limited response to all standard techniques.

Who are these men? They are the dangerous acting out people, the sexual psychopath, the murdering rapist, the highly aggressive, dangerous sexual aberrationists, the murdering man in Atlanta a few years ago who picked up black boys, abused them sexually and killed them. Then there was the man in Illinois who buried fifty people under his house. The "Son of Sam" is another, who shot to death young women as they sat with their dates in a parked car. These men look perfectly normal and act as we do, but they are difficult to detect because they are extremely clever at covering up their crimes.

These psychopaths come from all segments of society and are mostly males. They account for the serious sex crimes and fortunately, they are extremely rare. The news media's affinity for sensationalism makes their numbers appear greater than they are.

What is happening in their brains? The center in the brain that controls aggression and the center that controls sexual excitement are situated close together, and there is a wiring defect in the brains of these individuals that is inherited. When both centers are aroused, an unhealthy aggression results, ending in rape, assault or murder. Both these areas are stimulated by the male hormone, testosterone, and as soon as a sexual arousal occurs, the murderous rape appears. A normal man is the opposite. He is tender, loving, desirous to please and gives and receives affection and support when sexually stimulated. There is much we don't yet understand, but we do know that it is an electrical problem in the brain.

What is the treatment? There is no effective treatment, every type has been tried. These men are impossible to treat with present knowledge because the circuitry in the brain cannot be changed. Nothing stops them from repeating their crimes, so they should be prevented from contact with others, which means lifelong incarceration or execution. Some are helped by lowering with drugs the levels of the male hormone, testosterone, which is associated with sexual desire and sexual aggression. One drug, MBA, requires weekly injections, another CPA, must be taken orally every day. There are few side effects but most patients are lax about taking their medicine.

The reaction of the public to the treatment of convicted sex offenders by these antiandrogen drugs is mixed. Some regard the penalty as not severe enough, others as too severe, and there are some who are suspicious of the treatment, as it allows the offender an escape from punishment.

Behavior modification has been tried but found wanting. It is effective for a short time, but once the influence of the therapist is removed, the effects of the therapy vanish. Outpatient techniques consist of recommending satisfaction through masturbation and training the offender in social skills to divert aggressive sexual needs into worthwhile channels. This may help in some cases and has some acceptance. Castration is a final answer, however, its constitutionality is questionable.

Psychodrama and analytic psychotherapies are for the most part ineffective, although some disagree. However, any type of psychotherapy would have to be intensive and long term, expensive and doubtful as to efficacy.

The larger group of sex offenders who are merely an annoyance are of both sexes, predominantly males. They use what might be called "aggressive seduction" techniques to entice the victims, such as sexual advances of a petty, moderately harmful nature but without long term results psychologically or physically. However, they are annoying and unpleasant.

Here is an example. A man came to the office for therapy with this story. Every day he goes looking for young boys. He learns their greatest interest is motorcycles, so he acquires one. He approaches a group of youngsters, picks out the most passive one, and asks if he would like to go for a ride. The boy usually agrees, and the patient "works him", as he put it. He takes him for more rides, then to his house, then to buy ice cream and so forth. Finally, when the time seems right, he asks: "Would you like me to make you feel good?" Then he masturbates him or performs oral sex. It is

astonishing how often these boys give in.
Is this harmful? The memory of a seduction of this kind is distasteful, and the old school claims it turns a boy homosexual. This is not true. A boy cannot be made homosexual; he is born that way. Such an experience doesn't help and it doesn't inhibit sexual growth, but it leaves an uncomfortable feeling about having allowed intimacies by a person of the same sex. Although self-esteem and self-image may be impaired in sensitive persons, mutually agreed upon sex play is usually innocuous.

Occurrences such as this are fairly common. Many people, women and men, are exposed to minor sexual deviations in school and other situations. An emotionally secure person suffers no lasting harm, and the experience can teach a young person about life. These sex offenders are not assaultive or dangerous. Many are business and professional people, politicians and workers in many fields.

What makes a man look for his sexual needs in these minor sexual perversions? The tendency is present in some men and a few women and is caused by wiring defects in the brain. In other words, the brain links sexual feelings to cause a person to do things we consider abnormal. These people have no control over their actions because the defects connect sexual feelings with the wrong action. This is built into their systems in ways we don't understand and the medical profession has found no treatment that can change them.

Are they caught and punished? Yes, many are caught and some are punished, but unfortunately, the great majority of sex offenders who are apprehended are not those involved in serious crimes. The public is apt to suspect them of being rapists and killers because they seem similar to the tiny group that is dangerous. Furthermore, they are not adept at evading detection and are easy to catch for they don't consider their actions wrong and neglect to hide them. However, most exhibitionists are caught, transvestites quite often and a few rapists and assaultive homosexuals sometimes. But these are not the most dangerous sexual offenders. The really dangerous criminal is extremely elusive.

The police go after minor offenders with great diligence, as they are easily identified, especially when a man resorts to petty crime, such as robbery, sexual exposure or unacceptable behavior in order to indulge his whims, and this gives the police the chance to find and arrest him.

As many as fifty sex crimes have been defined, so laws are necessary to protect against acts that range from those of little consequence to murder. It is essential to report them all, as they may help identify real criminals before more victims are injured.

There are about fifty thousand sex offense arrests each year. As a rule, the least intelligent and the least paranoid offender is the most obvious. The police can easily pick up the petty thief but fail to nab the professional burglar, and they are adept at arresting the man exposing himself on the subway, who is more of a nuisance than a menace, but have trouble catching up with the hardened sexual criminal, such as the repeat rapist.

Laws have been on the books for several hundred years but are seldom used to restrict sexuality. If they were enforced, most of the population would be classified as sexually deviant. Vaginal intercourse between husband and wife and self-masturbation in private are the only legal forms. All other practices are illegal. Be grateful that the arm of the law does not extend into the bedroom! We need laws, however, for protection against the abuses of sexual freedom out of deference to those who support them for moral reasons. So society quietly condones the pleasures of sexual conduct between consenting adults. Sexuality is a personal matter and not to be interfered with provided it remains mutually acceptable and not illegal or injurious.

What about the child abuse we hear so much about? Some people who like to work with children are immature, and there is a homosexual tendency in a small number. For example, some Scout leaders, private school teachers and camp counselors are often homosexuals. They love to be with young people and see them in the shower, swimming in the nude and having close contact with them. These persons should not be classified as sex offenders, for they rarely assault or abuse children and fill a real need in their roles. The dangerous sex offender is a different breed. Nevertheless, all group teachers and directors of children's activities should be regularly checked for illegal sexual actions.

What causes this type of behavior? As already explained, these minor sexual aberrations result from wiring defects in the brain, so when sexual feelings are aroused, the person is directed into abnormal instead of normal outlets. Sexual responses are programmed in the brain, influenced by events, knowledge and experience and directed in a way that is prearranged. It is believed that this took place at puberty with the normal introduction of male or female hormones, or because of a brain injury or birth injury or a genetically developed birth connection. Thus, the brain directs the person into a sexual activity that he favors.

These individuals feel no guilt nor do they want to change, as their sexual appetite is satisfied in whatever manner they are programmed by the faulty wiring.

What are these mild perversions? Some have interest in children or objects or parts of the body. Another is aroused by women's clothing or seeing people's feet. Wearing female clothing or makeup satisfies others. None are homosexual, nor would they engage in homosexual activity. Otherwise, they are perfectly normal and make good husbands and fathers. A few men indulge in intercourse with animals, which is considered a perversion but is relatively harmless to another person. Notwithstanding, these perverted activities upset society, and the press exaggerates and over publicizes them.

Help is available for this group as they are amenable to psychiatric treatment. However, most don't seek it because they are not aware that what they do is unusual. An occasional man will pose this question to the doctor: "Why do I have this desire to touch little girls and boys? Why do I always reach out for them?" These men have more thinking personalities, and are therefore more sensitive. But they don't think through what they do and are not really interested in changing, probably because they get sexual satisfaction that way.

In their later years, these men become less able to function heterosexually, and turn more to their perversions, such as masturbating with female clothing or with the object of their desire, and have little or no normal sex. This causes their wives great unhappiness.

The public takes a profound interest in the sexual proclivities of others, but no matter how you try, you cannot force anyone against his or her will to engage in sexual perversions. Some men find vicarious satisfaction observing a man dressed in women's clothes or makeup on his face, but you can't get him to do it himself. There is a deep natural resistance in most of us which we don't yet understand.

THE RAPIST

Rape is the most common crime of the sex offender. Sometimes it is preceded or followed by robbery and occasionally ends in murder. The law stipulates that intercourse with a woman by force is first degree rape. From the legal standpoint, the aggressor is always a man, and he may either act on impulse or plan the crime. Rape in the second degree is intercourse with a girl who is a minor, even though she consents.

Some rapists are men with a child's emotional development, which tends to make them feel sexually inadequate. They usually choose young girls, because they are easier to take advantage of, are inexperienced and weaker. If a grown woman is assaulted, she may submit when she cannot fight him off, so as not to get hurt. A court of law usually accepts her version of what took place. Some women are stimulated sexually by indulging in fantasies of rape, but they would find the real thing quite different.

Why do men rape? A man doesn't rape for sex, for often he has a wife or girl friend with whom he has sex regularly. He doesn't hate women, but sincerely believes that women are attracted to him, and that his victim finds him desirable. His excitement intensifies if she resists him. Recently, a pioneer in research concluded that rape is "an act of aggression and violence, motivated primarily by power or anger, rather than by sexuality".

Rapists usually commit other crimes as well, but rape gives them the chance to dominate another human being. It is also a challenge and a thrill to get away with the forbidden and this builds them up. Some rapists engage in homosexual acts, even though they are primarily heterosexual. They indulge in "one night stands" or latch on to an older man of means and end up blackmailing him.

What is in the mind of the rapist? In past years, rape was considered the result of an irrepressible sexual urge triggered by the way a woman dressed, looked or acted. More recently it has been accepted as simply violence against women visited on them through sex. The latest research suggests that rapes are many faceted and comprised of varying degrees of aggression, violence and high sexual drive. It also shows that there are only a small number of rapists who are driven by hatred of women or sadistic fantasies. Most of them are men who succumb to an impulsive urge as the opportunity presents itself. This often translates into date rape. Fifteen percent of a recent survey of a large number of college women claim that they were raped and fifty-six percent were raped on a date.

Many studies and surveys have been made in order to determine why men rape, but none has come up with a definitive answer that relates to all cases. Some convicted rapists are aroused by violent sex and inhibited by consenting sex. So research by many scientists determines unequivocally that there are many reasons why a man rapes and many factors involved. Several are angry at a certain woman, hatred of women, abuse as a child, the opportunity presented by a date, a desire to rob, a woman alone in her home and the excessive aggressive nature of some men. And these reasons are not all.

The question arises, why? As we know now, thinking and doing is activated by the brain with its myriads of circuits. We believe that there is defective circuitry in the brain of the man who rapes. A faulty wiring directs the sexual drive into the wrong direction which appears to be a congenital defect. These instincts start very early, continue through life and seem not to be changed by therapy, jail, punishment or restrictions.

In a study of men, over fifty percent were inhibited from having sexual arousal when hearing of a sexual encounter in which the woman was forced to have sex and experienced distress and pain. When hearing of consenting lovemaking, they were not, which indicates that violence inhibits sexual arousal in the ordinary man. Convicted rapists demonstrate the opposite reaction. They were aroused by violent sex and inhibited by consenting sex. However, further experiments showed that the ordinary man, when angry at a woman or by a woman, was sexually aroused.

Some rapists have an uncontrollable compulsion to rape. Recently, a method of treatment is proving to he effective for those who want help in preventing their crimes. A daily dose of a female hormone called Depo-Provera reduces the sex drive while not causing impotence. Surgical castration is effective, but rarely used because it is irreversible and constitutionally questionable.

The specter of sexual harassment recently came into the media's focus during the confirmation hearings of Judge Clarence Thomas. Undoubtedly, there are degrees of this activity ranging from the casual comment offered to a woman in the work place on how she is dressed or looks to an invitation to lunch or dinner to an attempt to embrace her. What happens next depends on the response of the woman to the overture, which some women regard as a compliment and others as insults. Any woman should be able to rebuff such unsolicited advances.

Rape, however, is far beyond sexual harassment and unfortunately the law does not recognize other forms of rape that can be as destructive as the criminal kind. Every day, boys are the victims of men, of older boys or girls of women. Girls are seduced by men, by women and older girls. Here are instances of child molesting.

The medical term for child molesting is *pedophilia*. It is taken from the Greek "love of children" and is due to defective wiring in the brain. Pedophilia is prevalent although there is no knowledge of the incidence in the population. It appears to be more common in men than in women. Heterosexual and homosexual men are guilty and some are partially or totally impotent, so the abuse may not be intercourse but a masturbatory type of handling of a young child of either sex, with display of the sex organs. Underlying this is fear of not being able to perform, and it may be that they are sexually satisfied only in this fashion.

When the molester is inadequate, retarded or sexually perverted and cannot find sexual satisfaction due to lack of finances and educational background, he takes advantage of children to satisfy his needs because they are weaker and gets his way through force, sometimes brutal force.

There is a mild form of molestation when homosexual males are attracted to Boy Scout outings. They like to camp out with boys and play at sports with them. They get vicarious pleasure watching the boys showering and swimming in the nude. Occasionally, they are active sexually, however, this doesn't cause homosexuality in the normal adolescent, but does produce resentment, and a tendency to disparage homosexuals in later years.

What causes pedophilia? Here again is the age long controversy of nature versus nurture. Researchers have identified the onset of these sexual perversions in children from three to eight years old as having been fostered by traumatic family and social experiences. No studies have been done to our knowledge on genetic influences, so there is no balance by which to make a judgment. So the decision on the cause of these practices will have to be determined by research. Therefore, the origin of pedophilia remains elusive.

Therapists who treat child molesters have learned from their patients that they knew during adolescence that they were interested sexually in young children. The fact is that they have always felt too ashamed to discuss what they considered a shocking and immoral act, but could not curb the compelling need until they were apprehended by the law.

Is there a treatment for this affliction? Yes, but there is no sure cure. Hormones and other drugs are used to suppress the abnormal sexual urge and may offer a guide to solving the problem. Progestin, a synthetic form of the hormone progesterone, and similar drugs are given to control the sexual urge and partially block the effect of male sex hormones. However, it does not interfere with normal male sexuality or act as castration. It is most effective with psychotherapy, so the patient can turn to normal sexual activity for satisfaction. Experience indicates that patients are able to reduce the medication eventually without returning to former deviant practices.

Pederasty is a worse crime. This means anal intercourse with a child, and is a form of molestation which can be extremely harmful, physically and psychologically. The offender swears the child to secrecy by threats of harm to him or her and the family. Unless he is caught, many of these crimes go unreported, for the sex offender is not a criminal until he is apprehended. His only concern is to be found out, for he has no conscience.

The current exposure by the press of the rash of sexual abuse of children by the staff in day care centers brings to light the enormity of this problem. It has existed as long as time, but always swept under the rug. Close to fifty thousand children are violated every year in this country. Sixty-two percent of the criminals who are caught are given probation and only seventeen percent are sent to prison for three years or less. Those who are released or on probation are repeaters, since they have no other means of sexual satisfaction.

Mothers have formed an organization to expose the laxity of the laws and permissiveness of the courts. We wish them luck.

The *exhibitionist* has a compulsive, uncontrollable urge to display his genitals to women. Only in this way can he attain sexual satisfaction. He is not interested in normal sexual relations, nor is he aroused by the sight of a female body. As he always performs before strangers, he can keep his secret from his family unless he is caught in the act. As the victim is involuntarily subjected to this perversion, it has an aggressive tone, but there is no sexual attack. Such men are passive in all other respects.

The exhibitionist is compelled to return to the scene of his last act, which shows he feels guilty and wants to be punished. He also has a tendency to report others like him to the police, which confirms these feelings. This perversion is seen in men who are small time criminals, and they rarely marry. Exhibitionism also plays a major role in male homosexuality of the flamboyant type.

The cause of exhibitionism is unknown, but appears to be an inborn defect of wiring in the brain. Such men are often impotent, which creates sexual anxiety and insecurity. By exposing his genitals, a man proves to himself and to women that he is potent, and wants to tell the world. However, he never exposes himself in his own home.

Voyeurism is the counterpart of exhibitionism. It is uncommon in women. The voyeur gets sexual gratification by observing the sexual organs and activities of others. You read of "Peeping Toms" in English literature, who went to elaborate lengths to look into a lady's bedroom window. This behavior ranges from secret viewing of a woman's body to watching performances of sexual acts in houses of prostitution and other illegal establishments. These shows usually consist of several forms of perversion, mostly sadistic, and may include intercourse with animals, as well as homosexual and perverted heterosexual and group sexual activities. The voyeur has an underlying fear of castration, and watching others perform assures him that this won't happen to him. He can never be sure, however, so his appetite is insatiable.

Voyeurism is not condoned in our society. Establishments that encourage it are subject to legal sanction and voyeurs are punished.

Frotteurism is a sexual disorder that occurs primarily in adolescent males and gradually diminishes through life. It consists of frequent, intense sexual urges and fantasies that compel the person to touch and rub against a woman who is a stranger and unsuspecting. It is called committing frontage and is usually done in a crowded place, such as the subway or a bus by rubbing his genitals against her thighs or buttocks. This arouses him and he fantasizes that the victim is in love with him.

The act is harmless but extremely unnerving and distasteful, however, it does not last long as the fritter soon moves away in order not to be detected.

Zoophilia, meaning love of animals, takes three forms, singly or in combination, and all are indulged in for sexual satisfaction. First, watching sexual acts between animals, second, using objects, such as fur or stuffed animals, for sexual arousal and gratification, and third, intercourse with animals, called bestiality or scooters, namely coitus with animals via the anus.

Zoophilia is most common among rural persons such as shepherds, who live in close contact with animals. Kinsey found that nine percent of rural males had practiced it at some time. In clinical experience, these people are rarely considered by psychiatrists as psychotic but they are usually of low intelligence. There are instances of women having intercourse with dogs, but this also is rare, however, women sometimes use dogs for cunnilingus.

Can perversions be treated? Although little research has been done, the consensus is that they are genetic, influenced by personality makeup and environmental circumstances. It is believed that they are basic patterns, not learned behavior. All methods of therapy have been tried, from castration and vasectomy to hypnosis, including the association technique whereby the therapist tries to shift the patient's focus in the right direction. None have been effective. Psychotherapy has proved to be the most successful, but of limited duration, and incarceration only exacerbates the condition.

V THE IMBALANCE OF THE DAMAGED BRAIN

PREAMBLE

The brain is easily injured. Its tissue is delicate and fragile. Damage comes from outside trauma and inside disease. Much is from self-abuse.

Injury from outside is mostly from accidents, the head hitting an object or being hit. It can he a fall or a birth injury, a car or freak accident. It can come from an assault. Whenever the head is badly hurt, damage to brain tissue may occur. Wearing seat belts in cars and helmets while cycling is insurance in preventing head injury in the majority of vehicle accidents.

Damage to the brain from inside comes from self-abuse or disease. Both are described in this chapter.

When the brain is injured, the differences between the predominant thinker and the predominant doer shows up clearly. The brain tries to compensate and both thinking and doing are affected, but in different ways. If there is damage to a thinking person's brain, the ability to think is disordered, how badly depends on the severity of the injury, but there is less disruption in doing activities and the patient can still talk and move about. Damage to the brain of a doing person results in acting out irrationally, but thinking is clear, as well as the ability to relate to other people. Thus, the basic personality determines how a person reacts to brain injury, and treatment is designed according to symptoms. Damage to the brain reduces self control, so a doing person's loss of control results in uncontrollable doing. The reverse is true of a thinking person, whose doing is less impulsive to begin with, so has less loss of control. The doing person responds to sedatives and anti-convulsions, the thinking person to anti-confusional medication.

Injury to the brain is a tragedy and disturbing to witness. You feel sadness and compassion for the victim, then you wonder what caused it.

The question is whether it was from self-indulgence, or through accident or no fault of the patient. If the latter, your compassion increases, if the former, you place the blame on lack of common sense.

In all cases, love, compassion and above all understanding is essential, for not only does this contribute to the patient's emotional well-being, but hastens recovery. However, compassion and permissiveness can be overdone, and discipline, or what some call "tough love" is in order, especially in circumstances that could bring harm to the person or to others, such as allowing an epileptic to drive the car. Firm, strict control conveys security, limiting activities offer a sense of caring. A tight rein shows love, unlimited permissiveness shows a lack of concern. In each case the formula differs, depending on the personalities of you and the patient. Your love and compassion will show the way.

When thinking of "brain damage", you may consider that someone with a severe head injury or a brain tumor may have a damaged brain, but that is not all. What about the alcoholic, the drug abuser, the epileptic and the mentally retarded? Do they have damaged brains too? Yes, they all do, and the symptoms are similar. So we will discuss them as one, and explain what to expect and how to cope.

There are outward and inward signs to brain injury, and it can produce many kinds of mental symptoms, which vary according to the severity of the injury and its location in the brain. The most common is instability of mood, with rapid swings into depression or to being over-excited. The patient is impulsive and has poor self control. He disputes simple truths, and displays sudden outbursts of anger. He is restless and irritable. So whether the damage is slight, severe, or in between, it causes him to be unpredictable in his actions and susceptible to injuring himself and others. This is increased if the injury is extreme. These are the most disturbing signs, but the patient has no control over what he does or says.

What the patient does is to him not as bothersome as what he feels. These are the inward signs. He realizes his brain doesn't work as it used to because he is confused and has difficulty concentrating and remembering. He never had trouble before handling his job or managing money, but he does now, and it is hard for him even to do simple arithmetic. This upsets him greatly, but others are more concerned with what he does. So we should sympathize with the way he feels, and try to understand.

If he is the wage earner it is natural for us to live in apprehension of what is going to happen next. For if his brain has become damaged he is no longer capable of handling money or of using the higher powers of thought, such as planning ahead how to provide for the family. Can he continue to support us or will he get into trouble by spending unwisely, or engaging in activities that are wasteful, foolish or even dangerous? In a nutshell, he cannot be depended upon because he is unpredictable. Although sympathetic, family members may resent that he is not himself, and sense a loss of control over their lives. This eventually turns into rejection.

As he slowly recovers, and this can take years, the question is will he ever be well again? Is it safe to let him drive the car? Let his doctor decide that. But the patient says: "I see no reason why I can't drive!" If denied, he lashes out in anger. So the family lives in trepidation and constant conflict or let him have his way, although realizing he is not responsible.

Injury to the brain often results in loss of insight. Here is an example. A patient said: "I am the same as I always was, although a little slower, but I can still drive and I want my license back." This patient had a severe stroke and is greatly impaired in judgment, although only slightly in physical coordination. Therefore, it would be extremely dangerous to him and to others for him to drive. He will not accept or try to understand his limitations, and when thwarted, becomes belligerent, angry, demanding, insulting and difficult to deal with. And when anger mounts up, it triggers reactions he cannot control, because the brain is not functioning normally. A trifling challenge, a slight insult or rejection and he reacts.

This is the patient with brain damage.

The past thirty years have produced a group of people who abuse drugs. Whether it be marijuana, cocaine, alcohol, heroin, barbiturates, hallucinogens, other narcotics or a combination, after a number of years, it catches up with them. Instead of being mellow, relaxed and content, sort of snowed, as they were at first, they are impulsive, angry intolerant and occasionally physically abusive to those they live and work with, even to casual acquaintances.

Long term drug abusers have many automobile accidents, caused by an accumulation of toxins (poisons) in the brain resulting in an imbalance of chemistry. Only time will tell what the effects are, twenty, thirty, forty years hence. There is bound to be substantially more injury to their brains than we now suspect.

It is the same with the children, but unlike them, many adults are the cause of their damaged brains through lack of safety measures, neglect of self, or abuse of the body. But not so the child. He did not ask for it and deserves the greatest consideration, gentleness, tolerance and understanding. He can't help it when he flies off the handle and throws things about and kicks the first person he runs into. However, he also needs discipline, for this is his security and shows that he is loved. Work with him with firmness interlaced with love and understanding. Sympathize with the outbursts, and explain that they are unacceptable, and that you will help him control them.

Brain injured children are avoided by their peers because they explode at the slightest provocation. They are unpredictable and have poor self control. They burst out in anger at the drop of a hat and can inflict physical

harm. It is understandable why other children shun them.
Tolerance with sympathetic tenderness is essential in handling all damaged brains.

16 ADDICTION

What does addiction mean? The dictionary says "addiction is the state of having yielded to or given in to a habit, practice or something that is habit forming, such as narcotics, to such an extent that its cessation causes severe trauma." And what is it we give in to? It can be almost anything, the craving for food, gambling, the pursuit of a sport beyond reason, even work. It can be a drug, an activity or the introduction into the body of a substance that changes the mood. The use of narcotics is discussed in this chapter and alcoholism in the next.

History is full of references to substances that relieve pain. In the seventh century B.C., the Assyrians used opium, and after them the Sumerians called the opium poppy "the plant of joy", evidence that it was also used for pleasure. Two hundred and fifty years ago, the pleasure provided by the opium poppy began to flourish in Europe and particularly in China. Opium smoking was a way of life for many Chinese.

A hundred years later, pure morphine was extracted from the opium poppy, and by 1900 medically induced dependence and opium smoking and ingestion was common in the United States. Almost twenty percent of the population became exposed. Early in the 20th century, restrictive laws were enacted and these drugs were allowed to be dispensed only by a doctor's prescription. Clinics where addicts could legally obtain morphine were closed when it was apparent that they had no intention of giving up their habit.

Narcotics that are based on opium are called opiates and are used as a sedative or to relieve pain. There are two groups, opium and its derivatives and opium synthetics. The derivatives are morphine, heroin and codeine, the synthetics are demerol and methadone. They are all addictive, heroin the most, then morphine, demerol, methadone and codeine.

Cocaine, a New World discovery, is derived from the leaves of the coca tree and was used to relieve pain and resist fatigue. Today it has become society's scourge.

There are other drugs besides opiates, of which marijuana, (cannabis sativa), originally known as hemp, is the one most commonly used. It is a fast growing weed and has an equally long history as a narcotic and medicine.

All these drugs are extremely dangerous for prolonged use.

The discovery of substances to treat disease is one of the greatest of scientific findings, and their benefits to mankind are unquestioned. Many, like penicillin, lithium and thorazine, were serendipitous discoveries, but man has long known that some give pleasurable sensations, and by the end of the nineteenth century chronic drug addiction was rampant. Strict laws were passed to control illegal traffic in drugs, but proved of little use in stemming the tide for monetary gain.

During the twentieth century, people continued to use drugs for pleasure, so they were divided into "drugs" and "medicines", i.e., those that are "mind altering" and those that are not. Several that are beneficial in the treatment of illness were banned by law. For example, even today a physician cannot prescribe heroin for the relief of pain from cancer, because of its tremendously addictive nature.

Why are drugs so bad? Any substance that gives sensation of wellbeing or escape, a blurring of feeling or sense of euphoria, or any kind of "high", has a potential for abuse, and abuse is increasing, especially among young people. Studies show that addiction to some drugs can damage the brain and the body permanently. These are irreversible, while the effects of others are not. A major concern is that drugs reduce motivation and performance, particularly in children and adolescents, and when marijuana is mixed with alcohol, there is a high incidence of automobile accidents in a group already at risk.

Until 1962, only two percent of the population used drugs and use by the young was virtually unknown. By 1977, sixty percent of fifteen to eighteen year olds had tried marijuana, twenty percent stronger substances like cocaine and the hallucinogens, and about thirty percent had used drugs that are available only by prescription. No doubt these percentages have risen.

Of the sixteen million marijuana users, seventeen percent are males twenty to twenty-four. Amphetamines, "uppers", are another drug young people take frequently. The use of cocaine, the most expensive, is increasing also, especially in the higher economic group.

What is drug addiction? It is defined as continued use of any substance that, when withdrawn, causes severe emotional and physical symptoms. Tolerance is when larger doses are needed to give the same effect. Dependence is when withdrawal symptoms appear if the drug is withheld. Then the person is addicted and begins the cycle anew, for he feels normal only when under the influence, and withdrawal gives great physical distress. Most addicts neglect their health, many to a point beyond repair, in order to use their drug of choice.

Who are the drug abusers? There are two categories, those who take drugs supplied by their physicians for medical reasons, and those who become addicted by choice, which is the larger group by far. The medically addicted are seldom heard of, nor do they collide with the law. The others become social outcasts, and often resort to crime to satisfy their craving, such as burglary, drug peddling, pimping, or prostitution. Most of the thieving is from friends and relatives.

Drug addicts prefer to mingle only with their own kind, and develop a cohesiveness based on the need to satisfy their appetite for drugs. Alcohol releases inhibitions, often causing aggressive behavior in the alcoholic, but many drugs, except cocaine and the amphetamines, reduce aggression and the tendency for violence. However, during withdrawal, addicts can be violent also, and criminal behavior is common when trying to obtain drugs or money for drugs. Recently, drug abuse has become a major problem, but control is difficult since society has become more tolerant of drug addiction.

Does personality play a part? Yes, very much so. The more doing personality is apt to be susceptible to addiction and is affected differently than the thinker. He looks for a mental high and will dare to try anything, from sniffing glue to snorting cocaine, little caring what he might do to his mind and body. He is not insane or mentally ill, but tends to act out irresponsibly and get into trouble.

A thinking person, however, is less likely to become involved in drugs, as he takes fewer risks with his body. Furthermore, he is more severely affected by substance abuse, as drugs conflict with the chemistry of his brain cells. Since thinkers have a more sensitive brain chemical balance, the introduction of addictive drugs causes a disturbance in mental function and is less satisfying. At times, he becomes disoriented and subject to hallucinations. Most thinkers, however, might get a little high but wouldn't dream of going beyond that. Many don't even want to take prescribed medication, whereas the doer is willing and even demanding of his doctor to give him drugs. Some people enjoy hallucinating and losing touch with reality. However, a thinking person may go into a psychotic state and not come out of it. This is why mind-altering drugs are so dangerous, especially to this type of personality.

If drugs are so dangerous, what is the tremendous allure? All addicts feel pleasure whatever the drug. However, the effects are temporary, so that is why addicts are so often driven to go to any lengths, even occasionally as far as murder, to obtain more drugs. There is no evidence that the addict's life is enhanced in any way by this abuse. On the contrary, in order to experience a high, there has to be an electro-chemical drain on the brain. How often can you do that without injuring brain cells? We don't know, but we do know that many people have been permanently impaired by drugs, and that addicts don't survive for long if they continue drug use.

To be addictive means that a person is prone to becoming a slave to a habit. But every person is not addictive. Many can never be, even though they try. They smoke cigarettes, drink a lot of coffee, indulge in marijuana or cocaine and enjoy their daily drinks of alcohol. When or if they decide to stop, they can without trouble or withdrawal symptoms.

Then, why does the addictive person find it so hard to stop? Most find it impossible without long term treatment. Some can but seldom stay off for long. Permanent cessation of drug use is frighteningly rare. Hopefully, research will come up with the gene, if any such exists. How else can we differentiate between those of us who are addictive and those who are not?

And there is the "oral" compulsion, a need to put something in the mouth: gum, food, a cigarette, candy or a sip of coffee or soda. A person can be oral without being addictive. However, addictive people are usually oral as well.

Are these genetic traits? We are not sure, but suspect they are, and research is on going to find the gene that is responsible.

Some people get away with addictions. There are people who drink heavily all their lives and remain productive and creative members of society. They are good spouses and parents. They are alcoholic in that they drink more than the average person, but are able to handle it.

This goes for the habitual marijuana smoker and the addicted cocaine user as well. They continue to work and make good money and are energetic, bright, capable, strong people. Most are more doing than thinking, but they have a lot of intellectual capacity which they use to advantage, and are able to balance their life-styles in a way the average person cannot. But how long will they continue to get away with this? Hard drug users are not found among the over fifty age group.

Notwithstanding, the long term outlook is not good. Scientific studies show that of a hundred people who take cocaine regularly, ninety-nine wind up permanently injured, ill or dead. Of those who smoke marijuana, seventy-five will be the same, and fifty alcoholics will be dead by age fifty and ninety-nine by sixty.

What drugs are involved in addiction? The most addictive is heroin, which is double morphine. Morphine is a white powder and is taken by mouth in powder, pill, capsule or cube form or by injection, called "mainlining." It is the most potent of all narcotics and the one most addicts prefer. However, it causes severe disability, for the user runs the risk of contracting AIDS, hepatitis or other infections from contaminated equipment, and is at high risk of overdose and death. He has no control over the strength or the purity of the drug. He is also in danger of becoming under nourished and, therefore,, vulnerable to infections and disease, because addicts are interested solely in the effects of the drug and waste little time or money on food or health care.

Morphine has one-half the strength of heroin, is light brown and taken by mouth in powder, pill, capsule or cube form or by injection. It has the same harmful side effects as heroin.

Cocaine is produced from the shrub cocos erxthroxylan. It is a white flake like substance. It falls into the amphetamine group, but is not like the psychotomimetics, the LSDs and other hallucinogens, which mimic psychoses. It is taken by sniffing but can be injected. Its effect is different than that of opium, and produces imaginary powers, frivolity, hilarity, lightheartedness, paranoia and a feeling of superiority.

Cocaine is popular. It induces euphoria and reduces fatigue. Ten million Americans have tried it and one to two million are current users. It is habituating and addictive. It has a powerful psychological and physical effect on the brain. Deaths due to toxic reactions have been reported, and nasal membrane damage is common in those who snort it. There were eight million visits to hospital emergency rooms in 1982 for reactions, double the number of the year before. No doubt there are many more today. Cocaine abuse is at epidemic dimensions at all levels of society, and is increasing among business executives.

Crack is a new form of cocaine and is taken by smoking. A study at a V.A. psychiatric hospital suggests that more than ten percent of its patients are freebasing cocaine. Some smoke it, some inject it, and others snort it. All use alcohol and marijuana as well. More fights with friends and between husbands and wives occurred when cocaine is freebased or injected than used intranasally. Sixty-eight percent developed tolerance and seventy-four withdrawal signs. More than half experienced psychotic symptoms, such as persecutory delusions and hallucinations.

The growing use of cocaine in its various forms has become widespread from the 1980s on. The addictive and physically damaging effects are poorly understood by the general public, but the medical history of the drug is very clear. It is terribly addictive and cannot be used for long without severe damage to the brain and body.

What are the effects? Cocaine can be detected in the urine three days after a dose. It causes euphoria and talkativeness, increases energy and alertness and lessens need for sleep. It decreases appetite and in some, heightens sexual awareness and athletic ability. Permanent effects are shakiness, rapid heart rate, agitation, and dilated pupils. Occasionally, a user may have an acute and unpredictable toxic response such as panic, delirium, confusion, extreme paranoia or violent behavior. One gram of pure cocaine can be lethal.

As abuse continues, the cocaine addict is more agitated, suspicious, has repeated motor disturbances and grinds his teeth. He also has severe insomnia and weight loss. A full blown paranoid psychosis can develop with hallucinations and delusions similar to schizophrenia, making diagnosis difficult. At this stage of heavy use, cocaine no longer gives euphoria, but the addict continues to dose himself in the hope of regaining its pleasant effects. When he comes "down", he experiences depression, fatigue, lethargy and excessive need for sleep. How long the "crash" lasts depends on the length of use and recovery may take hours, weeks or months.

A few years ago, there was a rumor that cocaine is not addictive. This is entirely false and is proved conclusively by observation and many studies.

Barbiturates are depressants, called "goof balls" on the street. They produce symptoms similar to those of alcohol. If someone acts drunk, and you don't detect the odor of alcohol, suspect barbiturates. They cause confusion, difficulty in thinking, impairment of judgment, and mood swings of elation and depression. They cause irritability and loss of control, fighting, weeping and so forth. The tongue is sluggish and the addict falls asleep easily and sometimes lapses into coma, which can end in death. They are more dangerous than alcohol because the patient doesn't vomit. He should be forced to stay awake and his stomach must be pumped out.

Barbiturate addiction has become less prevalent during the last ten years and often goes unrecognized. Habitual use ends in irreversible deterioration of the brain cells. Anyone who uses barbiturates for sleep can become addicted, and stopping results in withdrawal symptoms. Often the alcoholic takes sleeping pills for insomnia, caused by the toxic effects of the alcohol. This combination can be fatal. An overdose of barbiturates, intentional or accidental is a leading cause of death from drug overdose.

Excessive use of depressants over a long period can result in physical and psychological dependence. Withdrawal leads to convulsions, especially with barbiturates, which may produce permanent disability or death. The combination of barbiturates with other depressants, particularly alcohol, increases the chance of death.

Amphetamines, "uppers", are taken to counteract the effect of barbiturates, "downers". The classic pattern to stimulate body and mind is to take uppers in the morning to get started. By nightfall, the subject is so keyed up that sleep is out of the question, so he takes a downer and the vicious cycle continues. Damage to the brain is irreversible. A person can be freed by stopping the amphetamines and gradually reducing the barbiturates. However, there are always withdrawal symptoms that are difficult to tolerate.

The glue sniffer is easy to detect, as the odor remains on his breath and clothes, and he often has a runny nose from irritation of the mucous membranes. The effect of glue sniffing resembles alcohol intoxication, and the sniffer experiences double vision, ringing in the ears, vivid dreams and hallucinations. Excessive use ends in stupor and unconsciousness.

Opium, derived from the opium poppy, is a dark brownish sticky substance with a heavy odor. A single dose is commonly known as a "toy", which can be eaten but is usually smoked in an opium pipe mixed with tobacco. Opium is addictive and highly dangerous. The user goes into a stupor and falls asleep or becomes unconscious.

The hallucinogens are LSD, lysergic acid diethyamide, peyote buttons from the peyote cactus, PCP, phencyclidine hydrochloride, and psilocybin from rare, toxic mushrooms. LSD and PCP are the most popular, the latter having a reputation as a "bad" drug without an antidote, yet many use it regularly, and it was associated with at least a hundred deaths in 1977 and more than four thousand emergency room visits.

LCD is taken with sugar cubes to mask its bad taste. It causes severe hallucinations, incoherent speech, vomiting, cold hands and feet, and a feeling of complete detachment. Instead of pleasure, it often induces a "bad trip", during which the subject experiences terrifying dreams and visions. Schizophrenia can be brought on in the susceptible person. Today, the use of hallucinogens is not as prevalent as in the 1960s.

The historical use of psychedelic drugs is intriguing. However, they have little therapeutic value. History tells us that the Vikings chose certain warriors as "berserkers" to whom they gave these drugs. Hence the word "berserk". When these men attacked, they inspired terror in the enemy. Their ferocity was such, as demonstrated by their irrational behavior, that they were impervious to the pain of wounds.

Marijuana, called "pot", "jive", "grass", "joints", "Mary Jane", "reefers", are all cannabis sativa, derived from a plant that grows wild and is also cultivated. The dried leaves are rolled in cigarette paper and smoked as a cigarette or in a water pipe. The smoke has a sweet smell. Marijuana can be baked in cookies, but requires a longer period for the effect, as only ten percent is absorbed into the blood stream through the intestines, whereas fifty percent goes directly to the brain when smoked.

Marijuana distorts thinking and intensifies whatever mood is present. It gives a false sense of power and a pleasant feeling of disorientation. The most common problem is loss of concentration and failure of motivation. These traits are intensified in those prone to anxiety and paranoia. Thinking personalities can become psychotic, which may be irreversible. Some people go into a rage and commit serious crimes, which may account for some of the murders today.

Research on marijuana is contradictory. One group claims: "Marijuana is not all that harmful", another: "It is very harmful". Studies by the New York Academy of Sciences in Costa Rica and Jamaica compared heavy users to non-users and found no significant differences between the two groups in a wide range of physical tests. The users were average Jamaicans who had smoked at least seven high-potency joints daily for eighteen years. This study is questioned by some, i.e., what sort of lives did these people lead? Were they workers or dependent loafers? Or were they constructive members of society functioning in an intelligent manner. This is not stated in the study.

The opposite viewpoint was reported in the Journal of the American Medical Association by a group from Columbia University. They concluded that marijuana interferes with cell division and damages lung tissue more than tobacco, with the damage persisting long after cessation of smoking. It reduces sperm count and decreases testosterone levels, causing birth defects. It also impairs brain function and judgment, and may trigger epileptic seizures.

NIDA, the National Institute on Drug Abuse, and ARF, the Addiction Research Foundation, its Canadian counterpart, first labeled research on marijuana "inconclusive". This drew so much criticism that they issued a consensus report: "Smoking marijuana irritates the throat and may cause bronchitis. It suppresses certain immune responses in the lungs, making users more susceptible to infection. It deposits fifty percent more tar in the lungs than one unfiltered cigarette, however, the quantity of smoke is less because few joints are smoked, but if cigarettes are smoked as well, the deposit is compounded, and increases the risk of cancer and heart attack.

What are the risks? Marijuana smoke contains as many carcinogenic hydrocarbons as tobacco smoke but lacks nicotine, a potent poison. It increases the heart rate, putting at risk those with heart disease. It reduces testosterone levels, sperm count and motility, and can cause gynecomastia, (swelling of the male breasts). It crosses the placenta and appears in the milk of nursing mothers. Marijuana impairs driving, especially in conjunction with alcohol. It can be detected in the blood eight days after use. Despite widely publicized findings that marijuana causes brain damage, later studies question this. However, it does diminish the ability to think and to concentrate.

Smoking marijuana is a risk. As knowledge grows, the dangers to mind and body make it unacceptable. Lungs bring fresh air into the body to aerate the cardiovascular system without which the body dies. Breathing smoke is equivalent to bringing a foreign substance into the lungs, and it remains there.

These dangers are not a deterrent to marijuana smokers. Many people consider it harmless, and it is used frequently at social gatherings of all ages. It can, however, be the worst of the drugs in that it leads to further experimentation. Although all marijuana smokers say they will never go to hard drugs, there are over 50,000 addicts in New York City who started with marijuana or other "soft" drugs. Countries where marijuana has been in use for a long time have severe penalties for its possession or sale.

Marijuana was banned by Congressional Act in 1937, but today researchers are rediscovering its therapeutic values. It is successful in treating the nausea and vomiting of chemotherapy for cancer, and it has recently been legalized in pill form for this purpose. In future, it may be used to treat asthma, epilepsy, glaucoma and depression, on which research is ongoing.

In the past, Queen Victoria's personal physician prescribed it for migraine headaches, the ancient Chinese used it as an anesthetic, and the Vietnamese to treat allergies, rheumatism and tapeworm. In the last century American doctors recommended it for eye strain, upset stomach, depression, nervous tension and epilepsy.

Less harmful drugs are also abused. Aspirin, although useful, can be overused. It causes bleeding of the stomach and can affect the entire vascular system. If given to children for fever from a viral illness, it may bring on the condition called Reyes syndrome, which can be fatal.

Caffeine also is a powerful nervous system stimulant with addictive potential, and is harmful in excess. Its content in various beverages is listed in Table 1, in milligrams.

There is habituation from caffeine. Surveys show that about twenty-five percent of coffee drinkers consume five or more cups a day, or five to six hundred milligrams of caffeine. Researchers have set 200 mgs. daily as the danger point above which a person is at risk of caffeinism, from which millions of Americans suffer, most without being aware of it. Symptoms are similar to those of anxiety neurosis, nervousness, irritability, tremulousness, muscle tension, insomnia, sensory disturbances, frequency of urination, loose stools, stomach upsets, extra heartbeats and palpitations.

Caffeine is especially bad for children and adolescents. They become irritable and have irregular heartbeats and insomnia when they drink coffee or colas with caffeine. Those with heart disease, ulcers, and other gastrointestinal problems are also at risk, as are people with anxiety and emotional conditions.

TABLE I

One 8 oz. cup		One pill	
Coffee, dripolated	146	Caffedrine	200
Coffee, percolated	110	Bivarin	200
Coffee, instant	66	No Doz	100
Tea bag, 5 mins. brew	46	Pre-mens Forte	100
Loose tea, 5 mins. brew	40	Aqua-ban	100
Tea bag, 1 min. brew	28	Vanquish	33
Cocoa	13	Empirin	32
		Midol	32
One 12 oz. can		Anacin	32
		Dristan	16
Coca Cola	65	Exedrin	65
Dr. Pepper	61		
Mountain Dew	55		
Diet Dr. Pepper	54	One bar	
Tab	49		
Pepsin-Cola	43	Chocolate	25
Diet RC	33		
Diet-Rite	32		

Women who are heavy coffee drinkers and have fibrocystic disease of the breast can reduce or reverse this by cutting out caffeine. Some studies show that there is an increased risk of heart disease in heavy coffee drinkers. Others find no connection.

If a pregnant woman drinks five or six cups of coffee a day or soda with caffeine, the risk of birth defects in her babies is increased. This can also cause complications of pregnancy, the possibility of spontaneous abortion, premature births and stillbirths.

What does caffeine do? It causes dramatic increases in blood pressure, muscle tension, and secretion of stomach acids. It reduces the amount of oxygen to the brain and increases the basal metabolic rate by which oxygen is utilized.

There are severe withdrawal symptoms from caffeine. The first sign is a throbbing headache, made worse by exercise. People treat these with Anacin or Exedrin, which they may not know contain caffeine, and the habit begins anew. Other symptoms are disorientation, constipation, nausea, depression and irritability. Tolerance builds up with heavy use, requiring increased quantities to achieve the desired effect. Ten grams of caffeine, the equivalent of eighty cups of coffee, can kill a healthy adult male, however, as it is excreted rapidly, deaths from caffeinism are virtually unknown. If caffeine were a newly synthesized drug, the Food and Drug Administration would probably not approve it. If licensed, it would be available only on a doctor's prescription, and there would be little need for caffeine, as it has no medical use except for rare cases of poisoning by CNS depressant.

Tranquilizers are prescribed for anxiety, stress, tension and nervousness. It is estimated that twenty million people swallow six million pills every day. The most popular is Valium, then Barbitol (the first barbiturate), Equanil and Miltown. In 1963, Valium was introduced by the Hoffman La Roche Company and took over most of the tranquilizer market as a result of intensive advertising. At one time, over fifteen percent of the population took Valium regularly.

Of all the substances doctors give patients, tranquilizers are the least harmful, but can be dangerous when used with alcohol or other drugs. They are not considered a threat to life and, although they are habituating, it is easier to stop this habit than to stop smoking, drinking or the use of narcotics. Tranquilizers are picked up by receptors in the brain, so the brain must produce something similar, a natural tranquilizer, perhaps the brain neurotransmitter, endorphin.

What do tranquilizers do? They are habit forming, and some individuals become addicted. There is little risk of overdose if taken alone, but many combine them with alcohol. Nine hundred deaths have been reported, undoubtedly a figure much lower than the actual. Dorothy Kilgallen died as a result, Karen Ann Quinlan remained in a coma for seven years before dying of pneumonia, and Betty Ford has publicized her problem with tranquilizers. In fact, hundreds of thousands of people are habituated, and most don't realize it.

Five to ten milligrams of a tranquilizer produce a calm mood, but larger doses induce drowsiness, loss of coordination and concentration, apathy and listlessness. Tolerance comes with daily use and can develop within a week. A few months can bring physical dependence and habituation, to some, addiction.

Habituation varies with the individual and can be determined by two tests, if tolerance requires larger and larger doses and by the severity of withdrawal symptoms. What are the withdrawal symptoms? They range from mild anxiety to epileptic seizures. There is restlessness, agitation and insomnia, with headaches, nausea and vomiting. Shaking, trembling, sweating, and body cramps may follow, along with skin reactions and a sensation that something is crawling over the skin. Medical care may be needed to effect total withdrawal as painlessly as possible, and the support of family and friends.

Why do people get hooked on drugs? Most cases of addiction to medicinal drugs can be traced to the physician. Some doctors give in to the impassioned pleas of patients to "Give me something to make me sleep" or "pep me up", and continue to renew prescriptions. If a doctor refuses, most drugs are available illegally.

There is a difference between habituation and addiction. The severity of withdrawal symptoms determines that difference. When they are severe enough to require medical care, it is addiction, when they are not, it is habituation. The same drug can be addictive to some and habituating to others.

After excessive use, alcoholism is an addiction, for withdrawal results in delirium tremens, requiring hospitalization. Caffeine is habituating, as is smoking tobacco and marijuana. Although a person is extremely uncomfortable when he stops, the symptoms are not debilitating or life-threatening. However, all opiates are addictive, as are their derivatives, and extremely injurious to the mind and body.

Nevertheless, habituation can be as harmful as addiction. It is as difficult to give up smoking or coffee as it is heroin or cocaine, and requires self-discipline and the strength to undergo a period of distress. Actually, it is the pleasure derived from the habit that the person is unwilling to give up.

Can we prevent drug abuse? This is difficult but should be attempted. Opinions differ on how, but there is agreement that we must change the social climate of youth. Present efforts focus on the mystery of drugs, which enhances the incentive to try them. The best way is to educate the young to the devastating effects of drugs and build up in their minds another image. Fortunately, most don't succumb, but the frightening experiences of peers can be a valuable deterrent to those who are attracted.

Young people are risk takers and experimenters, and their doing side is much stronger than in later life, when the thinking side is more developed, and less apt to give in to temptation. A more effective course is to promote sports and physical fitness, the desirability of preserving optimal health, and the responsibility of each individual to maintain it.

The community helps also through youth groups, churches and other organizations that enlist the interests of young people by opening communications between adolescents and giving out information not covered at home or in school. Programs on health promotion and disease prevention are long overdue and the health community can enter the picture and play an important part.

The Government's role is to stop illegal entry of drugs into the country. Organized crime is responsible for most of this and the government is doing the best it can against tremendous odds, not the least of which is the laundering of narcotics money.

The multiple approaches by sea and air and the long border between the United States and Mexico make it impossible to police every area. The vast network of the transportation system provides easy access for quantities of drugs to enter the country concealed in various ways in large and small shipments and hidden in or on a person or in his luggage. These drugs are worth thousands of dollars on the street, with risk being the rationale for the high prices.

What is the incidence of drug abuse? A recent study by the National Institute of Mental Health of five thousand persons in three major cities reveals that seven percent of the population abuses or is dependent on alcohol or drugs. That translates to 6.9 million people who are addicted to hard drugs and 19.7 to alcohol.

How do you cope living with an addictive person? Living with an addictive person is one of the most difficult tasks in life, whether it be to alcohol, drugs, nicotine, gambling, food or other substance or activity, The most important consideration is to take care of yourself, the children, the household, the finances. These are your responsibilities, it is not your job to fight the illness. That is his or her responsibility.

Self-help groups are everywhere and there are "Anonymous" groups starting with Alcoholic Anonymous, Cocaine Anonymous, Gamblers Anonymous and so forth. The policy they advocate and treatment they use are the most effective there is. Doctors are not successful in treating most addictions. They admit this, and give credit to these groups by referring patients.

So, the policy that works is to step over the body of the alcoholic as he lies on the floor in a drunken stupor and go on about your business. Eventually, he will "hit bottom". Then, let him find his own help. It may be difficult to follow this advice. Most people can't, and try to keep the offending drug away from the patient, avoid parties and contacts and whatever puts temptation in his way. Forget it, you are only wasting your time.

Observation by professionals over many years shows that about two-thirds of addictive persons tend to have more doing personalities, while the remaining one-third are thinkers. For instance, the doing alcoholic is sneakier and more apt to become aggressive and unpleasant and, when he is drinking, he may get into a fight or be assaultive. The thinking alcoholic gives an appearance of being upright and conscientious but, as his alcoholism progresses, he becomes psychotic, paranoid and very disturbed.

We don't know the nature of the brain chemistry that causes addiction, although extensive research is ongoing. It tends to follow inherited patterns, as it is not explained by other rational measures. Though it may not start in younger years, the tendency is there at birth, and can begin at any age.

Many addicts, especially the thinkers, can handle their addiction and cover up for quite a while by compensating with intelligence, business or other activity. In the case of an alcoholic, the ability to detoxify decreases as he reaches middle age, and he becomes a victim of what for another would be social drinking. Such people are perennially cheerful, positive, happy and hide their illness very cleverly. Members of the family can't stop them. They are going to have their liquor.

We ask the same question over and over. Why? It is difficult to have compassion for a person who ostensibly caused his own brain damage through abuse of drugs. Why did he do it? Perhaps he is not aware of what it can do to him. He does not know the harm is irreversible. Perhaps he inherited an addictive personality, had the money, was tempted and got hooked, or his personality is conducive to recklessness and taking chances. It doesn't help him or you to be intolerant, so forbearance and sympathy make him feel better and you too. But you cannot cure the person who is addicted. He has to help himself.

How do you treat drug addiction? Like the alcoholic, the drug addict must first admit that he has a problem and wants to submit to treatment, and therapy is based on the premise that addicts are on a day-to-day survival. Methadone can be used as a less harmful substitute for heroin and sometimes for cocaine as well, however, its effectiveness is questionable, and it is postulated that once an addict always an addict.

Methods of treatment are detoxification, psychotherapy, group therapy and counseling, drug treatment, electrical stimulation, acupuncture and methadone maintenance treatment. However, this last is not instituted until other forms have failed. The latest therapies that are in progress in clinical trials are certain drugs, auricular acupuncture, Cutaneous Electrical Stimulation (CES), Neuroelectro Therapy (NET), and the National Institute for Drug Abuse (NIDA) is experimenting on various substances in animal and human studies that are in progress.

Methadone clinics are strictly regulated by federal and state laws and serve about 80,000 patients who come every day for a dose. The clinics also supply vocational training, group therapy and social services. Sixty-five to ninety percent of patients stay in methadone treatment for more than a year, however, some drink heavily and abuse non-opiate drugs, but they commit fewer crimes and more often maintain family life and a job. After withdrawal from methadone, the results are not clear. Studies range from twelve to eighty-three percent of abstinence after one to three years.

Today, group living in specialized centers called therapeutic communities (T.C.) is the treatment for the addict. The objective is to render the person drug free, and is based on the philosophy of Alcoholics Anonymous. Although alcoholics can "dry out" within four weeks, it takes the cocaine addict longer and he must remain in this controlled environment for a year or two, depending on the progress achieved after one year. These T.C.s are administered and staffed in great part by recovered addicts who are dedicated to helping their fellow sufferers. The results of this treatment are inconclusive as the experience is short-lived.

In the sixties and seventies, therapeutic communities were populated mostly by heroin addicts. From the early 1980s on, most were persons addicted to cocaine. The long term effectiveness of this rehabilitation therapy depends largely on the social milieu into which the patient is returned and the attitude of the free society in this country that openly condones the self-administration of legal substances such as tobacco and alcohol, and closes its mind to interfering in the recreational use of illegal drugs by the individual as being "none of my business", although not necessarily countenancing it.

How effective are the other types of treatment? Realistic figures are not available, but experts maintain that addicts who admit their problem and submit to therapy can be cured, provided they remain in follow-up care for a year after completing a program. The active involvement of family and friends is essential and there are numerous self-help groups based on Alcoholics Anonymous to which they can belong, such as Cocaine Anonymous, Gamblers Anonymous and Synanon for the family and friends of the addict. Regardless, the results are not encouraging, as it is estimated that one percent or less remain abstinent and most addicts do not submit to treatment. In regard to cocaine, rewards from treatment are short-lived because of the nature of the drug, as adverse effects appear sooner than with alcohol or other drugs.

Cocaine addiction has evolved within one decade from a virtually nonexistent problem to an extremely complex disorder, in which are involved environmental, psychological, and neurophysiological factors. Cocaine, within hours to a few days of use transforms a normal, healthy person into a craving maniac with an uncontrollable compulsion to experience again the same euphoria which is short-lasting and soon demands another dose.

Cocaine has indeed become the scourge of society, not only in the United States but in every country in the world except Australia. That is probably because its location is out of the mainstream of the drug traffickers. China and Japan have succeeded in controlling it. European countries have curtailed its use to a larger extent than the United States but still have a way to go. On the other hand, China and Japan have instituted strict interdiction laws whereby there is life imprisonment for traffickers and seizure of their assets, and compulsory hospitalization and rehabilitation of users. This has resulted in a reduction of the number of arrests of users from 2200 to one hundred in four years. In Japan, in 1973, the Ministry of Health stated that "To become a narcotic addict is to commit an anti-social act." This statement still stands and the people respect and adhere to it.

A study by a New Haven drug program finds that seventy percent of opiate addicts have a psychiatric disorder. Over two-thirds complain of depression, twenty-seven percent are antisocial personalities and fourteen percent alcoholic. Whether these problems were the cause of the effect of opiate abuse is not known. The suicide rate of these addicts is five times that of their age group, excluding overdoses which may have been deliberate.

The record shows that one-half of the addicts in the United States had been arrested before they took to drugs, so criminals are apt to become opiate users. The other half commit crimes to pay for the habit, and when both groups become dependent, their criminal activity increases and the crimes become more serious.

A study of addicts in Miami shows that they commit three hundred and seventy-five crimes a year. There are several types, one lives by stealing and selling drugs, another holds a regular job while feeding his addiction by theft and drug sales, and another lives on welfare, obtaining drugs from friends and relatives.

The addicted woman has different problems. She suffers from depression more than a man, and many have small children and live on welfare. If she continues her addiction while pregnant, her baby is affected and presents serious symptoms after birth. The addicted mother is not equipped to deal with a baby, much less a sick baby, so society has to rescue the child at great expense and the diversion of care from other patients.

Why do people take drugs? A person takes a drug for the pleasant effect he experiences which is caused by a change in the brain's chemistry. We are learning more about how drugs alter brain function, and have named certain structures in the brain the "reward system". This system is not well defined, but studies suggest that alcohol and cocaine, although of disparate chemical structures, stimulate the reward system.

This reward system is primitive and powerful, so primitive it has the same system as the rat, and so powerful that a rat as well as a man will abuse a substance until he dies. Apparently, stimulating the brain's reward system brings extraordinary surges of euphoria. This leads some to hypothesize that hedonism may be the biological basis of addiction.

Recent and ongoing research is finding out more and more about addiction, especially to alcohol and cocaine, the most prevalent in our society, and researchers are paving the way to unraveling the enigma of addictiveness. Why are some people addictive and others not? Why are some addictive not only to one substance but to two or several? Why is it that the person addicted to a drug may also be addicted to food or gambling or only to food? Scientists are concentrating on substance abuse, however, addictiveness per se is not being researched adequately.

What causes addiction? Some scientists hold that stimulating the reward system is the key, but not all agree. It is interesting that while humans can become addicted to alcohol, cocaine and heroin, monkeys become addicted to cocaine more easily than to heroin and rats are easily addicted to both, but it is difficult to get them addicted to alcohol.

It is not possible to give a precise definition of addiction. The American Psychiatric Association states that it is dependence and abuse of a substance, but dependence and abuse don't always go together. For instance, cancer patients take morphine to relieve pain but do not become dependent. The brain pharmacology of rats is similar to that of humans, so research was done on rats in order to determine what the reward system consists of. This indicates that when a small portion of the brain is removed, the rat no longer strives to get to the cocaine, probably because it no longer gives him the pleasure he worked so hard to obtain. Such research cannot be done on humans, but it is hypothesized that the process of drug addiction may be similar. Further research may determine one way or the other.

Are the genes involved? There is a good deal of data on genetic factors in alcohol abuse, and research has long established that alcoholism is genetic. Now, investigators are working on the predictability of alcoholism in the offspring of alcoholics. On the other hand, there is no evidence of a genetic predisposition to cocaine abuse so far. Research is still in the early stages. The effects of the two drugs are different, however, and this has been demonstrated in animals and humans. Although scientists don't agree about the effects on the reward system of alcohol and cocaine, it is well substantiated that cocaine acts directly on it, producing intense pleasure immediately, whereas alcohol takes a longer time, indicating that the effect is indirect. Moreover, it takes longer to become dependent on alcohol than on cocaine.

What is happening in the brain? Studies with rats by several researchers propose that the phenomenon of a repeated urge to take cocaine takes place in the prefrontal cortex of the brain, and that the nucleus accumbens of the rat brain sustains the desire. They find a strong correlation between the ability of cocaine to bind to the dopamine re-uptake sites and its ability to cause monkeys to take the drug repeatedly.

Between binges, the cocaine addict suffers a grim and cheerless state that indicates the lack of a sufficient supply of dopamine in the brain. This fits in with the hypothesis of a decreased amount of dopamine, as does a different theory that cocaine abuse causes sensitivity of dopamine receptors on nerve cells that release it, so that dopamine is working overtime, causing its depletion. Both could account for the craving for cocaine. Cocaine also stops the uptake of serotonin and noradrenaline into brain neurons, as well as affecting dopamine transport, and is a powerful constrictor of blood vessels. This can cause a heart attack, which may also occur with the use of phencyclidine, or "angel dust", as it binds to the same site as cocaine.

Unlike cocaine which is a stimulant, alcohol is a depressant, psychologically and physiologically, and over indulgence can lead to stupor. In this regard, researchers are studying the addictive aspects to a particular neurotransmitter system in the brain. Several appear to be affected, gamma-immunobutyric acid (GABA), serotonin, dopamine, noradrenaline, somastatin, acetylcholine and vasopressin. New studies show that alcohol may directly simulate the brain reward system by causing dopamine to be released from the nucleus accruement.

Because alcohol interacts with several neurotransmitter systems, one cannot predict which one it will choose or what the effects will be. The acute effects may be different from the chronic effects, and one drink may differ from a weaker or a stronger one. The physiological effects of alcohol are the reduction of anxiety, sedation, motor incoordination and the removal of inhibitions. Substantiating evidence shows that these effects are related to the GABA system in the brain, the same on which tranquilizers also act. Some research indicates that alcohol may cause the brain to produce morphine, thus explaining alcohol addiction. The counterpart of morphine in the brain is the endorphins.

In conclusion, the cause of drug addiction will not lie solely in how the brain is affected. Researchers stress that the chronic taking of a drug results in a powerful conditioning which is extremely difficult to reverse. Also to be considered, especially with alcohol, is that the consumption of spirits and wine is universally condoned, and has long been accepted as a social amenity. This makes it conducive for those who have an innate predisposition to become addicted. These environmental factors should not be discounted.

17 ALCOHOLISM

Man has drunk and enjoyed beer, wine and spirits throughout recorded history, and excessive use is described as far back as in the Old Testament in the story of Noah, (Genesis 9.21). The use of alcohol originated in religious and ceremonial customs, and problem drinking is still rare among groups where this is the primary use, for example, the Orthodox Jews. The Moslem religion forbids wine, and some Christian sects consider any but sacramental use a sin.

Over the years, the growth of social intercourse led to the use of alcohol to promote relaxation and kindle congeniality. Until recently, drunkenness was considered a social blunder or a moral weakness, hut today it is recognized as a medical illness. And alcoholism is not a crime or a moral weakness. It is a disease, an addictive illness and a major health problem. Its complexity defies clear-cut description, but we do know that it is related to brain chemistry.

What causes it? Some scientists believe there is a biochemical or metabolic factor, others a personality element, and still others theorize that it is a combination of these two or an inherited tendency. Studies show that the alcoholic has biochemical and metabolic elements not present in the social drinker, and psychological testing points to a personality element as well. There may be an inborn predisposition that grows with the individual, or an inherited tendency, both irreversible and encouraged by social customs.

Inheritance is definitely implicated and research shows that there is a strong influence believed to be a genetic defect of metabolism in the handling of carbohydrates. Alcoholism recurs in families, though not necessarily in every member or generation. Some believe the tendency to become alcoholic is peculiar to certain ethnic groups which have more alcoholism than others. But this theory is weak, for there is no group that is free and none totally alcoholic.

What are the other genetic causes? Some people suffer from inherited adrenal cortical insufficiency, whereby the body calls for a substance that stimulates the adrenal glands. Alcohol does this and men in their forties and fifties are likely to show this insufficiency. If they are already drinkers, they drink more.

The most exciting clue to the cause and possible cure of alcoholism came to light within the past two years. Researchers at two universities and the National Institute of Drug Abuse discovered a gene that makes a protein in the brain that latches on to a dopamine receptor which has several variations. One of these they call the A1 version, and it is this version that is present at autopsy in the brains of seventy percent of alcoholics and only twenty percent of non-alcoholics.

Subsequently, this A1 version of the gene was found in the DNA of fifty percent of living alcoholics and twenty percent in nonalcoholics, a significantly higher percentage. Another group corroborated these findings.

Although this research is not yet conclusive, some scientists speculate that two or more factors combined with the A1 version and not the A1 version alone may be responsible for alcoholism. So a broadened scope of the research is planned to study the inheritance of alcoholism by combing through the genetic material of up to one hundred families to determine whether others along with the A1 version are responsible.

In addition, researchers have identified one of two genes that occupy a spot on chromosome 11 and direct the key function of dopamine receptors on brain cells. There is evidence that this gene is linked to severe alcoholism with medical complications and that the gene may be responsible for intensifying the severity of alcoholism rather than causing alcoholism per se by disturbing dopamine transmission. Dopamine is an important chemical messenger that helps to regulate pleasure seeking behavior. Although these findings remain controversial, they have been confirmed by others and six investigators are now conducting a large study of six hundred alcoholics and their family members. It is conceded, however, that the environment also plays a significant part in producing this disorder.

Since the first writing of this chapter, a late update in the hunt for the alcoholism gene comes from the results of a twin study just published on three hundred and fifty-six identical and fraternal twins. This shows marked genetic heritability for alcoholism among males who have symptoms of alcoholism before the age of twenty, but only a slight influence in women and late onset men. These findings dispel to a large extent the current belief that inheritance is a major factor in the cause of alcoholism. Researchers may have overlooked the significant influence the environment and social customs have on the desire of anyone to begin to drink. Therefore, there is little evidence of heritability for most alcoholics.

Predisposition to alcoholism seems to be passed on more often through the female. It is well known that women produce offspring with a high incidence of genetic malformations, low birth weight, and other disabilities if they drink during pregnancy. Therefore, alcoholic women should not use alcohol during pregnancy.

Is personality involved in alcoholism? Yes. The alcoholic is physically and emotionally ill, and has a biochemical imbalance that makes him unstable and insecure. His failure to recognize how much he drinks is a symptom of the illness. Scientists report that levels of anxiety, tension and resentment are measurable in the blood and drop to lower levels upon the ingestion of alcohol. This fosters another excuse to drink.

Some alcoholics do everything to excess, work play, sex, exercise, sports. Many are very dependent and often attach themselves to a strong personality, then wear him or her out by their demands. They can be selfish and manipulative and tend toward destructiveness. They are often immature, and blame those close to them or circumstances for their inadequacies. They are apt to have passive/aggressive personalities and deal with problems the easy way or violently. An emotional crisis may precipitate excessive drinking but, when problems are resolved, the drinking stops. However, the alcoholic can always find an excuse for his addictive behavior.

Alcoholism causes massive denial. The alcoholic denies his drinking and is evasive about other things as well. He is deceptive in regard to alcohol and goes to great lengths to hide it and make sure that it is always available. This compulsion spills over into other areas of his life, and even before he starts to drink, the trait shows up in childhood and young manhood.

Young people will try anything, so it is understandable if they experiment with alcohol. It is "the thing to do", so they give it a try. It is socially acceptable, not illegal and easy to get, as most homes stock it. Some want to find out how much they can drink and those who can tolerate the most are at risk for alcoholism, for excessive toleration is a warning. The individual is not aware of this, in fact, he is proud of the fact that he can drink others under the table.

Thinkers and doers react differently. Studies show that thirty percent of alcoholics are thinkers and seventy percent doers. Occasionally, a thinking person reacts violently to heavy drinking, and shows signs of schizophrenia. When the alcohol wears off, he returns to normal. The protein metabolism of the alcoholic is already out of kilter, and if a thinker adds another drug like cocaine, marijuana, heroin or LSD, he will become even more disturbed, leaning toward the psychotic side. People with doing personalities go on benders and commit crimes while under the influence of alcohol. This releases normally inhibited drives, causing aggressive behavior. If another drug is added, they act out even more, and veer toward the psychopathic side. Then, they may commit serious crimes.

Children are often the targets because they cannot fight back. Hence, the child abuse you hear so much about. An eighteen month old boy is dead on arrival at a hospital, having been beaten by his stepfather; a three year old girl drowns in the bath tub, having been held under the water by her mother's boy friend because she wouldn't stop crying, an eight month old baby is brought to the emergency room by his mother with bruises on his face and body and a dislocated shoulder. His mother mistreated him because he would not obey.

These are serious problems, and are usually committed by alcoholics and drug addicts who have excessive doing personalities and are inclined to brutality. They lose all restraint and control under the influence of the drug and take out their frustrations on the weakest, the child. Their lack of conscience lets them attack anyone, but they are careful not to go after someone who might be able to fight back. Today, the easy availability of guns compounds the issue and more heinous crimes are committed.

Is the chemistry of the body involved? Yes. In experimental animals, Vitamin B deficiency creates a craving for alcohol and slows down the digestion of sugars and starches. Genetic defects in enzymes produces the same craving. Thiamine, Vitamin B1, is also connected with the metabolism of sugars and starches. A deficiency of Thiamine, creates the need for alcohol as a sugar substitute.

Recent studies also implicate the endorphins, which are chemicals manufactured in the brain that kill pain. Morphine is actually synthetic endorphin. Alcohol stimulates the production of endorphins by a change in the brain's metabolism which is not well understood. When endorphins are low, a person who cannot produce enough endorphins, drinks excessively to get an anesthetic effect. The social drinker feels no need of this, thus, the alcoholic becomes dependent on alcohol just as the drug addict becomes dependent on morphine or heroin, which is double morphine, or two morphine molecules linked together.

Research in the treatment of alcoholism is promising. Endorphins are produced in small amounts in the brain and are extremely potent. It has been synthesized for use in research but is inordinately expensive. So far, results on alcoholism are encouraging, as well as on several mental disorders. The Food and Drug Administration is in the process of approving the use of endorphins in clinical trials on people. It is hoped that alcoholism will one day be treated successfully with endorphins, but at this writing they are unavailable anywhere in the world.

It is our theory that the endorphins will prove to be the controlling factor in alcoholism and all addictions. Observation by doctors and scientists over the years clearly indicates that what addictive people really want is pleasure. This is proven in animals. Monkeys trained with alcohol, cocaine, heroin and so forth go back to it for pleasure and get hooked on the need for pleasure. It appears that once an alcoholic takes a sip of alcohol he cannot stop, is totally helpless and has to drink until he is sodden. Something triggers this sequence and we must find the key to what does it. Long term drinking results in tolerance, so more is needed to produce the same effect. Alcohol is absorbed quickly through the stomach and intestinal tract, more rapidly in alcoholics, and travels through the liver to the brain immediately via the blood stream, reaching its highest level within half an hour to two hours. Eating slows down this process, but the alcoholic often skips meals, however, he may gain weight because alcohol has empty calories that add weight without giving nourishment.

Alcohol is eliminated by oxidation, most of which takes place in the liver, a lesser amount in the brain, and the remainder in the kidneys, skin, lungs and intestines. It destroys the liver by overloading it beyond its capacity to oxidize, resulting in cirrhosis, a life threatening disease.

Social drinkers say they like the taste of liquor, it relaxes them and helps them be at ease. Alcoholics usually can't tell you why they drink, they don't necessarily like the taste. Alcohol works on the central nervous system as a depressant, the first few drinks give a feeling of pleasure, because the brain's sensitivities are already depressed in alcoholics. Sustained drinking eventually produces depression, physical coordination suffers and enjoyment gives way to fatigue, and in extreme cases, to coma.

Although alcoholism is an illness, it was and still is not regarded as such. Historically, although not condoned, alcoholism is accepted in men. For reasons unknown, it is not as prevalent in women, but is more severe physically and psychologically. A woman is criticized for drinking too much and she knows this. She is not supposed to indulge herself in this manner, and the double standard by which she is judged makes her feel ashamed and guilty. So it is very difficult for her to go for help.

Most people know when they have had enough to drink, but not all stop at this point. However, there is no clear line between the heavy social drinker and the problem drinker, and the length of time it takes to become an alcoholic varies with the individual, as well as the degree to which he is impaired. From a psychiatric standpoint, a person who drinks enough to interfere with his job, his family and his social life is an alcoholic.

Here is an example. A man was a heavy social drinker until his forties when he began to drink more, especially on weekends. Many a Monday, he skipped work and stayed in bed all day. His wife knew he put away a fifth of whiskey each night and more over the weekend. Occasionally, he abused her, but never admitted it. When she tried to keep the liquor from him, he hid it on closet shelves and behind books. An alcoholic goes to great lengths to have a drink available and can outwit the most determined opponent. So the alcoholic continues to drink until he is so ill hospitalization is necessary and his problem is exposed, or some disaster occurs that confirms his condition.

What are the signs of alcoholism? First is intoxication, frequent blackouts and denial that drinking is affecting performance. Then delirium tremens, hallucinations and epileptic seizures, in that order. These are extreme and serious conditions, and potentially fatal if not treated promptly.

In delirium tremens the body jerks and trembles and the mind is confused with hallucinations, disorientation and blackouts. The hallucinations are mostly visual, sometimes auditory, and take the form of insects, animals or menacing shapes so terrifying that patients have been known to jump out of windows or run through the streets in terror. D.T.s is an acute psychotic, organic psychosis with hysteria. These patients can die from exhaustion and dehydration. They require massive amounts of fluid by intravenous injection and are known to rip the IVs out.

A rare disorder called pathological intoxication results from a sensitivity to alcohol and causes explosive reactions to as little as two or three drinks. The patient remembers nothing until later, when on recovering he hears about his outrageous behavior.

Alcoholism occasionally results from brain injury, which has important legal connotations called "legal insanity", whereby the accused cannot be held responsible for a crime. Only recently has this become recognized, and it is difficult to persuade the courts to judge such cases fairly. Electroencephalographic tracings show that brain injury related alcoholism is similar to a certain type of epilepsy, except that in epilepsy the syndrome is triggered by convulsions. Alcohol can precipitate seizures in epileptics, so they should not drink.

Some psychotics and psychopathics are also alcoholic. There are two serious psychiatric conditions which are extremely difficult to deal with. They are the schizoaffective alcoholic, who also has a thinking and mood disorder, and the psychopathic alcoholic, who has a predominantly doing criminal personality. The majority are more doing personalities, as alcoholism is more prevalent, but not limited, to that type. Depression is deepened by alcohol, which is also a depressant, and the two react upon one another to exacerbate both.

Mental illness and alcoholism, including substance abuse, tends to run in families, so the family history is important. However, it should not be considered a strong contributory factor in making the diagnosis but more of a supportive element. Moreover, it has been established that there is a link between alcoholism and certain psychiatric disorders.

We know that alcoholics experience depression, for alcohol gives an immediate lift, then a lowering of mood. A large number of these patients recover from their depressive symptoms soon after entering a treatment center where there is no access to alcohol. However, a small group continues to have persistent depression and develop physical symptoms, which often are associated with chemical depression (endogenous). These patients can be treated successfully with drugs, and this often results in the cessation of drinking.

Patients with manic depressive illness frequently abuse alcohol, especially when they are high in the manic phase. What they will not accept is that drugs of any kind (except their prescribed medication) worsen their chances for the stabilization of their disorder.

Patients with schizophrenia are difficult to assess, especially as to whether their symptoms are the cause of their illness or the abuse of alcohol, so it is important to withhold medication during alcohol withdrawal until thorough assessment is completed. This involves a physical and a psychiatric examination, drug history, family history, including lifestyle, aims in life and medical and spiritual background.

The differential diagnosis of patients with a mental disorder is particularly difficult, and the type of medication must be carefully reviewed, a regimen decided upon and adhered to.

How is alcoholism treated? While there is no guarantee, there are several methods in use and others being developed, and there is Alcoholics Anonymous.

The initial treatment is a twenty-eight day inpatient rehabilitation. This is now being reviewed and local outpatient versus inpatient therapy is being scientifically assessed. There are many advantages to outpatient therapy, as the patient can live at home and stay on the job. This alternative also offers an incentive to an individual to accept therapy at an earlier stage of involvement with alcohol. Many patients consider the twenty-eight day inpatient sojourn too long and too confining, and AA not enough.

Hundreds of studies have been conducted in the past ten years through clinical trials and the consensus is that "there is no single initial treatment approach that is effective for all persons with alcohol problems". The challenge for the doctor is to decide the best and the most effective, safe and economical treatment for each patient. It is implicit that some patients require inpatient treatment, such as those who are suicidal or homicidal, at risk for severe withdrawal symptoms or unable to avoid drugs in an intensive out-patient setting.

The most important factor for the physician is to remember that the cause of alcoholism is unknown, and therefore the medical care is unknown. Thus, self help is the best treatment at the present time.

Psychotherapy. If the patient admits he is an alcoholic and wants to stop drinking, psychotherapy is the most helpful treatment, but requires a thorough physical examination and psychiatric evaluation. The doctor may begin by prescribing vitamin and mineral supplements, particularly niacin and thiamine. He may also prescribe sedatives, tranquilizers and drugs to prevent delirium tremens. Then he tries to find out why the patient drinks and how he feels about himself under the influence of alcohol. Most patients lack motivation, however, and don't return after the first visit.

Many patients do best in group therapy where they can share and compare feelings and experiences. This gives them support and is less expensive than individual therapy. However, some patients are incapable of sharing their problems in a group and require one-on-one sessions.

Emetic drugs. These drugs condition the patient against alcohol. The one most frequently used is antabuse, which produces an aversion to drinking. It is taken daily, after which one sip of alcohol makes the patient violently ill, with nausea, vomiting, flushing, and palpitations of the heart. This is conditioned therapy and is successful in those who submit to it, but it requires psychotherapy as well, in order to prevent relapses.

Alcoholics Anonymous. This world wide organization is known for helping countless alcoholics and is dedicated to assisting all who want to join. It maintains that alcoholism is an allergic reaction to alcohol, which can never be safely taken in any form. Studies so far have not proved this, but it is supported by much observation. Members of AA are available on a twenty-four hour basis for those who are trying to stop drinking.

Former alcoholics founded the fellowship many years ago and groups meet daily in five thousand locations in the United States alone. These meetings are a form of group therapy at which members tell the story of their experiences. Doctors often refer patients to AA, which has made progress where the best of physicians and psychiatrists have failed. Thinking persons are least responsive to AA, as they are repulsed by the group approach and prefer individual counseling.

Acupuncture. Relatively new in the United States is treatment of alcoholism by auricular acupuncture. Although used in China for five thousand years for many medical conditions, acupuncture is beginning to be accepted in the Western world. The method is cheap, simple and has no side effects, and counseling, either group or individual, is helpful to the patient while undergoing acupuncture treatment.

The site that is targeted for alcoholism is the ear, hence auricular acupuncture. Tiny needles are inserted under the skin of each ear and left there for about twenty minutes. The frequency and number of the treatments vary with each individual, but it is reported that the results are encouraging and seem to replace the desire to drink. Patients usually obtain relief quickly, within ten to fifteen minutes. They feel better, have a better appetite and a renewed sense of well-being.

TENS, NET and *CES.* Another form of therapy for alcohol and drug abuse is transcutaneous electrical nerve stimulation. TENS is used extensively in physical therapy to promote healing of injuries and to ease pain. Other forms of the same basic method are neuroelectro therapy, NET, and cutaneous electrical stimulation, CES. Success in the treatment of alcoholism and drug addiction with these methods is mixed. Many studies are taking place, and time will deliver the verdict of their usefulness.

Naltrexone. Another form of treatment is with the drug naltrexone, which is used to treat heroin overdose. Current studies show that it helps alcoholics stop drinking by lessening the pleasurable effects of alcohol, and the relapse rates of those treated are cut in half. The drug has few side effects and, in conjunction with psychotherapy, it appears to help alcoholics over the hump of the first three to six months after they stop drinking.

So far, a small study demonstrated that persons on naltrexone occasionally took a drink, but stopped at one or two, while those on a placebo lost control and went back to heavy drinking. The drug seemed to lessen the pleasurable high of drinking. Larger and longer studies are in progress.

It is theorized by many scientists that alcoholics have a deficiency of endorphins in the brain. These morphine like brain chemicals seem to alleviate this deficiency. Naltrexone may diminish the craving for alcohol by neutralizing its effect.

Because of the enormity of the problem of alcoholism in this country, and indeed, throughout the western world, there is a good deal of work being done by medical researchers, pharmaceutical companies and the National Institutes of Health on the treatment, cause and cure of alcoholism.

The major barrier to treatment and the biggest hurdle for the alcoholic is to admit that he or she has a problem. Some go on for years, slowly deteriorating physically and mentally. Family and friends are brushed aside, unable to help. It usually takes a shock like the onset of delirium tremens, the loss of a job or an accident to bring the alcoholic to acknowledge his addiction. Alcoholics Anonymous calls this "reaching the bottom". Once the alcoholic wants help, there is hope.

The fatality toll of alcoholism in terms of the general population is devastating. One-half the 55,000 deaths from motor vehicle accidents each year are related to drunken driving. In addition are the injuries that are uncountable. According to the Committee on Tests for Intoxication of the National 'Safety Council, the safe blood alcohol level is up to fifty milligrams percent concentration. A percentage of 150 mg. is the level at which one should not drive. The body oxidizes alcohol at the rate of half an ounce of eighty proof spirits an hour. Thus, it takes four hours to eliminate the effects of a two ounce drink.

Alcoholism breaks up marriages, alienates friends, and affects every member of the family. It is the most frustrating condition in medical practice, and doctors dread treating alcoholics, for there is little they can do for them. Statistics show that one-third of patients don't return after the first visit, another third after the first month, and at the end of a year, ten to twenty-five percent are still in treatment.

It is estimated that alcoholism costs society many billions of dollars per year in wages, crimes, accidents, hospital and medical care, and the cost of incarceration. What cannot be counted is the lost creativity, productivity and the tragedy when family life is disrupted.

Alcoholism is a major social and public health problem. It is estimated that one-third of the population uses alcohol excessively, although not all are alcoholics. It is impossible to determine the number, a vague yardstick is twelve percent of hospital admissions. The uncountables are those treated privately or never treated, the incarcerated and those who join Alcoholics Anonymous.

What are the results of treating the alcoholic? Can he or she go back to social drinking? There have been many studies on this, yet no therapy has proved permanently successful. Two decades ago, an angry dispute on this issue took place between psychotherapists. One group maintained that this is possible and another, including Alcoholics Anonymous, holds that loss of control of the use of alcohol is inevitable in the alcoholic. This controversy prompted a famous study that reported a successful return to moderate drinking after undergoing intensive therapy including practice in controlled drinking. After two years, the patients who learned controlled drinking had fewer alcohol problems and after ten, nine were still alcoholic, six were abstinent and four were dead.

After twenty-five years, a consensus holds that controlled drinking seems to be an option only for alcohol abusers who are not severely dependent and have few drinking problems. However, although they have a choice of when to drink, they must observe rigid rules. So the answers to this issue are still inconclusive.

The outcome of even more recent research clearly demonstrates that alcoholism is not always a progressive disease. Ten to fifteen percent of men and two to four percent of women have a serious problem at some time in their lives. After a plea for help, ten to twenty percent never relapse.

So the efficacy of the treatment of alcoholism is still up in the air. These are established facts: those who faithfully attend a clinic or AA are more likely to recover and those who have stable marriages and jobs are more apt to attend. We also know that most recoveries from alcoholism are not necessarily due to therapy, as forty percent have recovered spontaneously and no more than ten percent of alcoholics are ever treated. Those individuals with a stable job and a satisfactory family life have the best chance of success and it appears that age, sex and duration of abuse matter less. Some patients develop an aversion to alcohol and have no desire to drink. This occurs voluntarily, after "hitting bottom" or after a serious life event, such as an accident, loss of a loved one or a job, or a religious or spiritual experience.

Withdrawal from alcohol can cause brain injury in cases of severe and long-standing abuse. After recovery, most patients say they don't want to drink any more but continue to drink. However, one of the authors has had several cases of brain injury who, after severe withdrawal reactions, had no desire to drink and over subsequent years did not do so.

Some cases of intolerable withdrawal symptoms develop a condition known as Korsakoff's psychosis which is one of the aftermaths of extreme alcoholism withdrawal. When not treated, these patients may go into this psychosis. Today, this is prevented by monitoring the nutrition and adding to the diet vitamin B1, zinc, other nutrients, tranquilizers and certain drugs.

Here is an example. Before we knew how to prevent this psychosis, Elizabeth, an extreme alcoholic, came to me. She was intelligent, a college graduate, married to a fine man and had three children. She would drink herself into a stupor, fall down and crawl on the floor to her room. This went on for years and finally she agreed to be hospitalized. Her withdrawal symptoms were unbearable and she went into Korsakoff's psychosis. She fabricated stories out of thin air about famous people paying homage to her and the expensive automobiles she drives and the wonderful places she visits. She was completely irrational for about six weeks when suddenly she became mentally sound again and proceeded to carry on with her life in a normal fashion. She never wanted to drink again and never did.

Amazingly, a number of patients have gone through these distressing episodes and come out of it not wanting to drink. All went on to live normal lives.

To date, there are no studies with a longer follow-up than ten years, and some authorities question the increasing involvement of profit making institutions in residential alcoholism rehabilitation. Moreover, most patients come for treatment when they are at a low point in their illness and usually improve for a while. It takes years of observation to check up on each patient and most are lost to follow up. Another factor is the variability of each case, and the extent and depth of the illness. They are all individual and there are many differences in length, depth and severity.

Alcoholism is a variable and complicated disease and every patient is unique. There is no way to measure this, therefore no way to diagnose the severity, tailor the therapy or prognosticate the outcome. Until research gives answers to these enigmas, therapy of the alcoholic will remain a hit and miss proposition. In the meantime, there are several avenues of approach, one of which may work, but first and foremost is the stipulation that the potential patient admit that he or she is an alcoholic and asks for help.

Tragically, the young succumb to alcohol all too frequently. Its use by young people is increasing. About eighty percent of eleven to seventeen year olds report having had a drink, more than half drink at least once a month nearly three percent drink daily. Since 1966, the number of high school students who became intoxicated at least once a month has more than doubled, from ten to over twenty percent. Nearly eighty percent of male high school seniors drink at least once a month and more than six percent drink daily. Youthful problem drinkers, those who get drunk at least once a month, number more than three million. That is twenty-five percent of the fourteen to seventeen year old age group. How many go on into life as alcoholics? We don't know.

18 BRAIN TUMORS

Occasionally, a growth forms within the brain and starts to enlarge. Any growth in this location is serious because, being inside the skull, it cannot expand without destroying brain tissue. Therefore, medical attention is imperative as soon as symptoms appear because a brain tumor is as dangerous as cancer.

What are the symptoms to look for? Early signs are vague headache, anxiety and an indefinable loss of the ability to think. Diagnosis is difficult at this stage, for brain tumors have symptoms similar to several mental disorders. As the tumor grows and destroys brain tissue, diagnosis is less difficult, for the senses and functions of the nervous system are affected, and there may be changes in vision, hearing, taste, smell, balance and coordination. By this time, a brain tumor should be suspected and a thorough neurological examination performed.

Are there different kinds of brain tumors? Yes, and its name describes its location. Most brain tumors are gliomas, so called because they start in the glial cells that support the neurons. They include astrocytomas, glioblastomas and medulloblastomas. There are meningiomata arising in the meninges and spinal cord, and there are pituitary tumors. Metastatic tumors (cancers) are extensions of cancers that start in the body and spread to the brain. Abscesses in the brain are rare, now that most infections are cured by antibiotics before they spread. Bruising and bleeding under the brain lining, called a subdural hematoma, may be found after injury to the head and can be easily treated if diagnosed accurately.

How do you make the diagnosis? This should be attempted as soon as possible as the earlier the diagnosis the better are the chances for successful treatment. New technology, such as ultrasound, CAT scans and magnetic resonance imaging (MRI) are helpful in making a correct diagnosis. Some brain tumors can be treated by surgery in the early stages, and radiation or chemotherapy can treat others.

Many tumors are treatable and curable, especially if they are treated early. The thought of surgery may be disturbing to some patients but hope and positive thinking with support and reassurance give the patient a sound emotional basis for recovery. We all should be aware of the power of the mind over the body for all illnesses, and that positive thinking is the way to go. The medical profession accepts this concept and promotes its use during all treating and healing processes.

19 EPILEPSY

Epilepsy is not a mental illness nor a sharply defined disease. For years experts have pondered how to describe it. The layman knows it as a convulsion or a "fit", when a person falls to the floor unconscious, foaming at the mouth and writhing as though in pain, although he does not feel pain. Actually, it is a group of electrical disturbances in the brain of varying degrees of severity.

What is the cause? Anything that goes wrong in the brain can produce an epileptic seizure. Most cases are congenital, meaning that something has gone awry in the brain of the fetus while in the womb. Some seizures follow an illness, or an accident or injury to the head. It could be a tumor, a degenerative condition, or a stroke, called a vascular accident. There are a number of cases called idiopathic, which means that the origin is unknown. That term is often applied to any illness of which the doctor is uncertain. Epileptic seizures are usually hereditary but some may be coupled with another factor. We will understand this condition better as we learn more about the brain.

Epilepsy may be compared to nearsightedness, (myopia), for which glasses must be worn in order to see at a distance. So, both these conditions must observe a few restrictions for safety, such as the myopic wearing glasses when driving or seeing clearly at a distance and the epileptic taking the prescribed medication. Although there is no cure, practically every case of epilepsy is treatable, some cases don't require it and those with treatment can he controlled so that patients lead normal lives.

What happens when an epileptic has an attack? There are two kinds.

The "grand mal", which is a major seizure, and the "petit mal", a minor. The grand mal affects the whole body and appears suddenly, or is preceded by warning signs of a few seconds to several days duration. These are changes in mood, a sensation of pressure in the abdomen or uncontrollable muscular movements. These signs give the more thinking person the chance to go to a safe place, lie down and wait for the seizure. The more doer will go about his business, pay no attention to the warning, and may end up in injuring himself by his fall.

When the convulsion is over, the person is relaxed and perspiring. Some get up, feeling perfectly well, others regain consciousness slowly and are confused, sleep for a few hours and awaken feeling fine and some wake up with a headache. There is no set pattern.

Some epileptics have a seizure while sleeping and awaken not knowing anything happened except for a headache or muscle aches, mental sluggishness and nausea. It is hard to determine if a convulsion occurred, unless the person fell out of bed, bit his tongue or showed some other sign. Grand mal seizures can occur in sequence, separated by minutes or an hour. In between, the mind is confused. If the attacks continue, the patient goes into a coma and the temperature rises to a high degree.

After a grand mal seizure, some patients experience what is called the "twilight state", which lasts from a few minutes to several days. Personality determines how a person reacts to this. The thinker is withdrawn, still and disoriented, the doer experiences anxiety, paranoid ideas and hallucinations. He may even become aggressive, attacking the first person he sees, or run into the street, a danger to himself and others. Some take long trips and find themselves in a strange place with no recollection of how they got there. In all these instances, epileptics don't remember what they do. This reaction is similar to the symptoms of amnesia and a person with amnesia may be an epileptic. So, twilight states are dangerous, regardless of personality, for the patient can have an accident or commit a crime and not remember.

The second type, the petit mal, occurs mostly in children. There is no warning and the child suddenly becomes unaware of his surroundings. He may hear what is being said but it makes no sense. He is quiet except for small movements, eyelids fluttering, facial muscles twitching and eyes staring. This lasts from one to forty seconds, then he feels perfectly well. Although his mind goes blank, he learns to hide the seizure and explains he was thinking of something.

The frequency of attacks varies in the individual and are often unnoticed. In a day, they can occur from once to a hundred times or take place days or weeks apart. A few cases are serious enough to disrupt daily life, but can be treated, however, these patients have to lead somewhat restricted lives.

Most petit mal seizures occur in children and it is understandable if they do poorly in school, for concentrated thinking is being interrupted over and over. Teachers may be critical or think the children are daydreaming, restless or just inattentive; some may be hyperactive, irritable and aggressive, according to personality makeup. If they are slow in school, they are thought to be mentally retarded. So, with these problems of behavior it is important to make a diagnosis. This is done by electroencephalography, which is similar to x-ray, and once a child is properly diagnosed and on the right medication, his performance improves.

Notwithstanding, young epileptics don't like the idea of staying on medication for the rest of their lives. They don't like the stigma attached to it, nor the possible side effects which might affect their ability to get good grades, and specifically, they have great concern about whether or not they will be allowed to drive a car.

Until now, it has been the practice to keep the young patient on medication and hope he would outgrow his epilepsy after two years, for there was always concern that the seizures would recur and with greater severity if it was withdrawn. Happily, there is recent research that gives a guideline on when medication can be stopped. Studies indicate that the peak time for recurrence of seizures is the ninth month after withdrawal of medication, and are less and less likely to recur by the end of the second year. Therefore, it is recommended that epileptics postpone driving until they have remained off medication successfully for at least two years.

Great strides are being made in the surgical treatment of epileptic newborns. At last there is a chance for them. These babies are born with uncontrollable brain seizures and are doomed to die of complications by the time they are ten years old. This story demonstrates what can be done.

When little Ryan was only eight hours old, his wee body convulsed and suffered up to twenty seizures in a day. Doctors treated him with anticonvulsant drugs to no avail, so surgery was recommended and performed when he was fifteen months old. Today at four years, he is evolving into a normal boy. He is developing more slowly than other youngsters and still requires physical therapy, but his seizures are under control and doctors expect his intellectual abilities will mature with no problems.

How was this miracle accomplished? First, PET (positron emission tomography) images of Ryan's brain were taken and, with the help of electroencephalography, the diseased area of his brain was located and mapped. Almost the entire left hemisphere was involved and removed at surgery. Ryan's case and many others currently being treated cast light on the remarkable recuperative powers of the young brain. PET imaging is proving to be a useful tool in charting the activity of the normal brain as well as that which is injured or diseased. This capability brought to light the tremendous burst of activity that occurs only during childhood which undoubtedly accounts for the tremendous learning powers of the young brain.

PET has proved to be the leader in brain imaging, and allows the doctor to outline in detail the location of seizure producing tissue. CT scans (computerized tomography) and MRI (magnetic resonance imaging) cannot do this. They can, however, identify other conditions not picked up by PET.

Researchers are now able to postulate that the resilience of the young brain appears to relate to the over-production of neurons during infancy and an excess amount of synapses making neuronal connections during childhood. Children up to eleven years have many more synapses in their brains than adults, although the child's brain is the same in size and weight.

The emotions play a part in the mind of the epileptic and can provoke a seizure on the slightest pretext. An adult convulses if he is frustrated and a child when he is scolded, or even when he is strenuously active. Fatigue is a contributing factor as well, from overexertion or insufficient sleep. Some patients have seizures only while asleep, and others can ward off an attack by keeping active or concentrating hard.

Strong stimulation of the senses of sight, hearing and touch will provoke a convulsion in some patients. Flickering light is a common cause, like the sun shimmering through the leaves of a tree moving in the breeze or the change of lighting on the television screen. Some patients have an attack when listening to certain kinds of music, and a few are able to bring on seizures by going into the sun or simply waving their hands before their eyes. They consider the sensation pleasurable.

A few people have a mental illness as well as epilepsy. These cases are more serious and patients are depressed and irritable hours or days before and sometimes after a seizure. During this period, they are sensitive, get into arguments and may turn to alcohol, depending on personality. The thinkers go to sleep, while the doers become paranoid and violent. Others are mentally dull for several days, conscious and oriented but with loss of initiative.

Outside influences can trigger a convulsion, not only in epileptics but other patients under certain conditions. Epileptics are apt to hyperventilate, as does everyone, when undergoing great stress. This fast, deep breathing induces alkalosis and may bring on a convulsion. The degree of stress determines the frequency of attacks. Insulin hypoglycemia, to which diabetics are prone, can precipitate a seizure, especially in the morning when blood sugar is low, and some suffer attacks when under the influence of alcohol. A small percentage of children, during the first five years of life, experience convulsions from high fever. They are not epileptic and recover completely.

Obviously, the complexities of epilepsy make a diagnosis difficult, but it can be done by electroencephalography. Small electrodes are placed on the head attached to an instrument which records the electrical activity in the brain. The patient feels nothing and the procedure is harmless.

What is the treatment for epilepsy? Treatment is medication, and brain tumor should be ruled out first before this is prescribed. Each case is different, and drugs are given singly or in combination and monitored carefully, as most have side effects. The dosage for children may have to be increased as the child grows, depending on the symptoms, but many children outgrow their epilepsy.

For the few who don't respond to medication, surgery should be considered. It has been successful but these cases should be handled by an experienced neurosurgeon in an institution where sophisticated equipment is available.

What is the epileptic's future? Once the condition is under control, the medication may be decreased, especially in children, but rarely discontinued. However, in the majority of adult cases, the psychological approach is of vast importance, and the epileptic must accept the life long need for medication.

Most epileptics are conscientious about taking medication, and the question often arises as to whether it is safe to drive. For unknown reasons, they have fewer seizures when under pressure, so they don't usually have one while driving. This is more apt to happen at a casual moment, while eating lunch or watching television. However, the possibility is a threat to the public, for to have a grand mal seizure on the highway would be disastrous.

Should epileptics marry? Yes, with the proviso that the facts are understood by both parties beforehand. Generally, these couples have happy marriages, and the question of children should be decided by them. However, if an epileptic is incapable of taking care of a child because of his or her handicap, or if both parents are afflicted, parenthood may not be wise.

The incidence of convulsive disorders in the population is .05 percent, in relatives of the epileptic, it is five times as high, or a risk of 1.5 percent of passing it on.

20 HEAD INJURY

Nothing is more valuable than your head. It is where the mind is located which directs everything you do. It is the seat of the senses and your personal identification. It is the source of thought, feeling and consciousness. No wonder you value it. You consider as serious the smallest injury, yet many people do little to protect it by neglecting to buckle seat belts and use helmets when cycling. Head injury causes 40,000 deaths every year from motor accidents. Hundreds of thousands are hospitalized, some never to walk again.

This chapter is limited to psychiatric disorders resulting from head injuries, which come under four classifications as follows.

Traumatic concussion. Concussion is the least severe and occurs from a fall or when the head is hit by an object, resulting in unconsciousness for a few seconds followed by return of consciousness. Sometimes there is loss of memory for several hours while the person functions normally. This type of injury is common in boxing. Boxers have been known to be knocked unconscious and get up and finish the round with no recollection of having done so. The length of unconsciousness is not a gauge of the severity of the injury which is more likely to be determined by the symptoms that follow. These are headaches, dizziness, apathy, irritability, fatigue, dullness of thought, inability to concentrate, difficulty in sleeping and in remembering, and intolerance to noise, light, heat, alcohol and exertion. Usually, symptoms disappear within six weeks with no residual effects. Repeated concussions may result in the "punch drunk" syndrome, a chronic state of confusion which is permanent and irreversible.

What is the treatment? Special nursing care is essential, with constant medical observation for complications. Sedation may be necessary for restless and anxious patients, but narcotics and stimulants should not be given. As a rule, psychiatric disturbances don't follow, but some cases require psychotherapy for anxiety and the shock from the long convalescence and its related costs.

Traumatic coma. Coma is when the patient is unconscious for a number of hours or days with other signs, such as lacerations, blood in the spinal fluid, changes in the electroencephalogram and neurological signs. X-rays may reveal a fracture of the skull, although this doesn't necessarily mean there is injury to the brain tissue, nor does its absence exclude brain injury.

As the patient slips in and out of coma, he is restless and disoriented when semi-conscious, so medical observation and nursing care are essential until he is out of danger.

Traumatic Delirium. A delirious patient is disoriented as to time, place and his identity. He is confused, restless and incoherent, and his senses are disordered. He is irritable and anxious and his speech is sporadic. Delirium often follows coma, and the personality dictates behavior. The more doing person wants to know why he is in the hospital and denies he had an accident. He may become belligerent and demand to be released. The thinker wanders about aimlessly, puzzled and questioning, and occasionally talks and acts as though he were at work. The person with paranoid tendencies thinks the hospital personnel are enemies and may attack them. These patients need careful watching to prevent harm to themselves and others.

Korsakoff's syndrome. Korsakoff's psychosis is usually associated with alcoholism but can also follow delirium, coma or twilight states due to brain injury. The symptoms are similar but there is no inflammation of the nerves of the arms and legs with severe pain. Patients give answers that make sense but have no bearing on the question asked. They are confused, disoriented, and can't remember recent events. They are on the defensive about their shortcomings, make up unbelievable stories with no basis in fact, and cannot do arithmetic or think in the abstract. When they recover, they don't recall the illness or what they said or did. A small percentage never get better.

How do you treat this? The treatment is the same as for concussion, and the outcome depends on the extent of injury to the brain, the patient's personality, nursing care, and the home environment after leaving the hospital. Although the body and central nervous system have tremendous powers of healing, recovery takes six weeks or more. Patients experience apathy, fatigue, headaches and dizziness which may persist for several years. Personality changes are rare with the exception of irritability, but occasionally they do take place. There are two groups, those with considerable brain damage and those with slight. The former includes serious mental deterioration, punch drunk fighters, and a chronic condition with convulsions.

Depending on the location of the damage, injury to the brain can produce many symptoms, from paralysis of part or all of the body to minor changes in personality and well-being. Only the psychiatric aspects are covered here.

The personality is a strong determinant of how a patient accepts his illness, reacts to it, and presents symptoms. Thinkers cherish their bodies and go to great lengths to care for them. A man is meticulous about his grooming and the clothes he wears. A woman makes up her face, has her hair done, and dresses tastefully and carefully. Doers may be clean and neat, but the man seems unaware that ties and lapels are narrow this year and the woman eschews make up and the latest fashions, preferring merely to be comfortable. Neither is concerned about how he or she looks.

Injury to the head has special emotional meaning to the thinking person, the doer doesn't place much importance on a facial or head wound.

What are the symptoms of chronic brain disorder? Patients are depressed and moody, and suffer from headaches, dizziness, irritability, fatigue and insomnia. They have poor tolerance for exertion, excitement, alcohol and sunlight. They are anxious and apprehensive and the paranoid element is accentuated, resulting in preoccupation with self. Special tests determine the extent of residual disability and may take several years to reach an answer.

There are other reactions to brain injury which we describe herewith.

Hysteria. As already explained, certain personalities tend to develop neuroses, and hysteria neurosis may follow injury to the head, although there is no organic brain damage. Hysteria can take many forms, from amnesia to paralysis, loss of sight or hearing to the "shakes" and the inability to walk. Careful examination rules out physical reasons and the hysteria is treated by psychotherapy.

Compensation neurosis. Some people have the motivation to develop a neurosis in order to receive maximum compensation for an injury. There are three kinds of cases, those for which there is no compensation who develop hysteria, those who get well regardless of whether they are paid, and a smaller group, who are not getting well, but suddenly recover upon receiving a large sum of money. There is a hint of malingering here which we go into later.

Traumatic neurosis. When there is extreme threat to life, as during an earthquake, a holocaust, or a major disaster at sea, terror can strike anyone. Some people panic, have excessive anxiety, and develop post traumatic stress syndrome. Most neuroses after head injury are hysterical, but nonhysterical forms can occur as well. Phobias become exaggerated, obsessional defenses and weird protective devices are created. All are treatable by psychotherapy.

Malingering, is a nonhysterical form. There are differences of opinion in compensation cases regarding malingering, as to whether the plaintiff is feigning symptoms or pretending they are worse than they are. It is generally agreed, however, that true malingering, the conscious, intentional imitation of symptoms, is rare, as it is easy to detect by physical, neurological and psychiatric examination. The malingerer exaggerates the symptoms to such an extent he soon finds it difficult to maintain them at a believable and realistic level.

However, the diagnosis of hysteria and malingering is not easy, and there is much disagreement in compensation cases. "Truth serum" and lie detectors are not reliable and can be bypassed by a clever doing personality.

Prior disease. Chronic disease, such as alcoholism, arteriosclerosis of the brain or, in the elderly, transient dizziness or slight stroke (TIA), can cause an accident. These conditions contribute to psychiatric disturbances after brain injury. Positive identification of the cause is difficult, especially in medicolegal cases.

Strokes, which are more fully explained in a later chapter, are not psychiatric conditions but we mention them as they are the most common of all brain injuries. A stroke is a hemorrhage from a burst blood vessel or a blood clot lodged in an artery (thrombosis), cutting off the flow of blood to the brain. This results in paralysis of part of the body, the extent of which depends on the severity of the circulatory failure and the location of the lesion. The primary causes are high blood pressure (hypertension), hardening of the arteries (arteriosclerosis), or diseases of the heart. A stroke can also be caused by an aneurysm, which is the ballooning of the wall of an artery that has become weakened. The thin wall ruptures causing hemorrhage into the brain.

21 MENTAL RETARDATION

Mental retardation is limited intellectual ability.

Before adolescence, the development of mental function is arrested, as well as the rate of natural growth, learning abilities and the emotional and social adjustment that are commensurate with the age of the individual.

The mentally retarded are also known as mentally deficient, feebleminded, mentally sub or abnormal, and mentally handicapped. It occurs in varying degrees of severity with a variety of clinical signs and causes. Intellectual impairment is the one common symptom. In some cases it is the only sign, in others there is distortion of facial features and/or physical abnormalities.

What is the incidence? There has never been a count made of retardates in the population. A conservative estimate is one percent, classified as to intellectual and social levels and physical manifestations.

Of the intellectual and social levels, which are determined by tests, five percent of all retardates are severely mentally defective, twenty percent moderately and seventy-five percent mildly. These are termed respectively, idiot, imbecile and moron. Intellectual ability is measured by the I.Q. or intelligence quotient, a test that compares the verbal and arithmetical skills of peer group members.

Idiots have a mental age of less than three years with an I.Q. of twenty, imbeciles range from three to seven years, with an I.Q. of from twenty to forty-nine, morons are mentally eight years old and up, with an I.Q. of fifty to fifty-nine, where the label of mentally deficient ends. The average I.Q. of the general population is one hundred, so people with I.Q.s between sixty and one hundred are not retarded but are less capable intellectually.

What is the cause of mental retardation? It arises from an inherited gene or is caused by disease or injury. The diagnosis is based on laboratory studies and clinical signs. There are five groups, genetic (inherited), Down's syndrome (mongolism), congenital, organic and head injury. In some cases more than one factor may combine to cause the defect.

There are two types, the physiological retardate, who has defective intelligence only, and the pathological, due to both genetic and environmental factors. The genetic are primary or endogenous (internal), and the environmental are secondary or exogenous (external) factors. The primary group is caused by dominant or recessive genes peculiar to the patient's family. The secondary group is caused by environmental factors, such as infection, poison, injury, malfunction of the endocrine glands or malnutrition. Thus, there are many causes of mental retardation, all differing in severity. The most handicapped retardates survive only a short time after birth, if not stillborn.

In making a diagnosis, a psychiatric examination is performed and tests are given to find the level of mental deficiency. Standard tests are the Terman-Binet, the Wecheler-Bellevue and Kuhlmann which determine I.Q. These tests are criticized, but no one has come up with improved methods, so they remain standard procedures and are used extensively. When the results of the tests are in, the doctor prescribes a regimen.

A treatment program is outlined consisting of training, teaching and guidance. This is carried out in sheltered workshops, where patients are taught tasks commensurate with their abilities and earn money to support themselves wholly or partially. There are special classes in schools for learning the necessary skills, there are vocational schools where patients can learn a trade, and clinics to teach proper social behavior. A large number of retardates adjust to their communities and become self-sufficient.

Research is blossoming in all areas of mental function, and new discoveries are being made to update treatment and offer broader potentials to retardates. Today, it is possible to define, treat, and sometimes prevent congenital and genetic diseases before or at birth. The metabolic diseases, such as phenylketonuria, glactosemia, and others that cause severe mental defects, can be easily prevented by elimination of the offending nutrient from the infant's diet.

In addition, science has developed methods of prevention and treatment. Amniocentesis detects defects in the fetus at an early stage, which can result in the correction of the condition by surgery within the uterus, or in severe cases, abortion.

Increased knowledge has led to obstetrical care that greatly reduces the incidence of many diseases of the newborn, such as jaundice, oxygen deprivation and injury to the brain during birth. While we don't yet know the causes of every case of mental retardation, modern technology is making tremendous strides in the education and training of retardates in areas that allow them to lead relatively normal lives and make a contribution to society.

Research has recently discovered the cause of the most common inherited mental retardation second only to Down's syndrome. It is a chromosomal defect, the direct diagnosis of which can be made by DNA analysis applied before birth. It is called the "fragile X syndrome" because the tip of the X chromosome has a tendency to break off. This discovery was made almost simultaneously by scientists in the United States, Australia, France and Holland.

Both males and females can inherit the fragile X chromosome, more males by about two to one, and pass it on to their children and grandchildren. Many carriers, however, do not show abnormal intellectual or behavioral symptoms. Eventually, it will be possible to detect the syndrome in people who have a family history of mental retardation and correct the faulty chromosome, so a couple can bear children without this fear. In the meantime, genetic counseling can be offered to couples with such a family history.

On the whole, the mentally retarded are happy people. They relate well to others and seldom complain. Their reactions are childlike in many respects, but they take direction well. They are gentle and anxious to please, and overjoyed when praised.

Notwithstanding, life poses problems for them. Many are physically handicapped as well, but as a rule adverse circumstances don't produce psychiatric disorders. Although faced with pressures, limitations, and sometimes abuse, most retardates accept their handicaps and lead relatively well adjusted lives.

Retardates have poor judgment, limited insight and have trouble making wise decisions, not only because they lack intelligence, but because they are impulsive. However, they are not apt to injure themselves or anyone else.

Families regard the retarded member in different lights. A retardate cannot function as others do at a level to maintain himself, nor can he discuss intellectual subjects. He has to be cared for, and some families use him as a scapegoat, call him the "black sheep" of the family and look down upon him. It makes some family members feel superior to poke fun and laugh at him and blame him for their faults and problems. Other families treat the retardate as an "eternal baby" to coddle and pamper, and many enjoy this. Some families put the retardate into an institution or a custodial facility.

Sex is the most difficult area for retardates and poses severe problems, as they are usually unaware of the consequences. Traditionally, they are sexually active, and if left on his or her own can get into trouble. Both the men and the women can be sexually aggressive in an unpleasant, primitive way, the men without restriction and the women foolishly. They are promiscuous and often develop disease, have illegitimate babies, and engage in prostitution. And when they have babies, they can't take proper care of them.

This has prompted the establishment of teams of professionals to educate the retarded about sex and the consequences thereof. They teach them in a primitive fashion, as one would a child, and the necessity to be careful, not permissive. These programs appear to be successful, although there are no studies as yet.

Retardates sometimes marry, and when they do it can be a problem if they have children. Will they be retarded too? Probably yes, but without knowing the cause of the parents' retardation, there is no sure prediction. Will they take care of the children? Historically, no. So here is a compounded problem and genetic counseling may be advisable, with the choice of sterilization or the use of contraceptives. But are the retardates capable of making these decisions and carrying through? This question remains unanswered.

Alcohol and drugs can turn the gentle, good-natured retardate into a dangerous man. He becomes assaultive, aggressive and irresponsible. He has no insight, is unable to cope and often resorts to crime to buy more drugs or alcohol.

Here is an example. The police brought a man to the emergency room of the hospital after a call by his family for help. He had become violent and hit his girl friend, his sister and the priest. He was a known retardate in the community, who drinks when he can get it and when he drinks, he hits people. It took five policemen to hold him. He broke every window in the emergency room and when finally incarcerated, tried to hang himself. He denied all this and blamed the police. This man cannot be kept in jail because he has not committed a crime, but he should be in a state hospital for life, away from society and under full time supervised care.

22 SENILE PSYCHOSIS AND ALZHEIMER'S DISEASE

Until recently, Alzheimer's disease was considered a separate type of mental failure because it appeared in relatively young persons in some cases, whereas senile dementia was associated with people of sixty-five years or older. So the two conditions were labeled pre-senile dementia and senile dementia.

In 1901, a Swiss psychiatrist and neuropathologist, Alois Alzheimer, published the case of a fifty-one year old woman, whose memory slowly deteriorated. When she died four years later, he found at autopsy that her brain was riddled with microscopic changes that he called plaques and tangles. Alzheimer was struck by the similarity of the symptoms of pre-senile and senile dementia and was the first to examine the brains of patients dying of the disease which was subsequentiy named for him. In comparing his findings with those of modern physicians, it is now understood that there is no difference between the two diseases.

There is general agreement, therefore, that pre-senile or Alzheimer's is identical to senile dementia. Another form of senility is due to hardening of the arteries, called cerebral arteriosclerotic psychosis, or multiple infarct dementia, described in the next chapter.

Alzheimer's disease or senile dementia is the result of structural alterations in the central nervous system (CNS). The CNS comprises the brain and spinal cord. These alterations may involve different sections of the brain, especially the nerve fibers and cells. This change in the brain tissue causes changes in brain function and in personality.

What are the symptoms? They are slow to develop, subtle and insidious. First, a slight failing in memory, than loss of ambition, taking longer to perform regular duties and trouble adapting to small changes in routine. A patient won't listen to new ideas, is less alert and quite inflexible. He borrows past experience to solve today's problems. It is evident that mental and physical function is slowing down.

For example, the person cannot remember recent events, such as what he had for breakfast, although he recalls his early life clearly. His mind sinks further into the past, reconstructing his childhood, since these memories are still vivid. These too slowly fade away until he cannot remember names or words or places he knows well. His speech becomes disorganized and he cannot finish sentences. He becomes unaware of his surroundings and loses his personality. Despite these defects, he may have a good appetite.

It is tragic to watch a brilliant mind deteriorate as the nerve cells in the brain disintegrate, resulting in increasing difficulty recognizing family members and finding his way about the house. As memory fails, he becomes confused and disoriented and cannot tell what day it is or what year. He forgets his age and time means nothing. Sometimes he gets out of bed in the middle of the night and prepares to go to the office, or leaves the house and wanders about, with no recollection of his name or where he lives.

At this stage, the patient sleeps off and on during the day and awakens utterly confused. He dreams a good deal and mixes up the dreams with events. This adds to his anxiety and disorientation, causing agitation and uncontrollable physical activity. For example, he may walk quickly up and down the room, go in and out of the house and move the furniture about endlessly.

The personality of the individual directs the slide into senility. The passive person enjoys his decline and likes to be pitied and taken care of, but he also blames himself and develops imaginary illnesses, negative thoughts, anxiety and depression. He may be suicidal at an early stage but as he loses function he becomes much less apt to injure himself.

The more paranoid person is irritable and irascible, and is convinced he is being robbed and cheated, and doesn't believe anything he is told. This often extends into the sexual area and a patient may accuse his elderly spouse of infidelity. Angry assaultiveness may occur, usually without provocation.

The more thinking person becomes deeply depressed, and the doer lashes out aggressively, quarreling with everyone. He swears, cheats at games and may commit petty thefts. All patients experience depression, alternating with euphoria and delusions of grandeur. The final stages sometimes produce visual or auditory hallucinations, the memory becomes worse and the confused mind fabricates impossible and absurd situations. This evaporates to a silent confusion in which the patient cannot feed himself or follow simple commands.

Some patients forget to bathe and neglect their personal hygiene. Diapers or catheters are necessary. As the disease progresses, they are feeble and uncertain on their feet. They lose weight and cannot take care of themselves. Mental disintegration quickens toward the end until communication is impossible. Usually, death occurs from an infection the debilitated body cannot fight off. Because of their severe symptoms, Alzheimer's patients need supervision in a hospital or mental institution, or full time

nursing care at home.
What is the incidence? Due to advances in medical science, people are living long enough to develop the diseases of old age, including Alzheimer's disease. The number of cases now is estimated at four million in the United States, affecting a few in their sixties and sixty percent by the age of eighty-five. It is the fourth cause of death in this country. However, many citizens retain all their faculties at an extreme age. Examples are Winston Churchill, Bernard Baruch, Dr. Albert Schweitzer, Karl Menninger and Herbert Hoover.

Bodily aging is inevitable as long as life exists. It is a natural process and starts when we attain full growth. The face develops lines, the skin takes on a wrinkled, crepey look, and the body fills out or thins. The impression is of deterioration, but this is not so. Many people maintain good physical and mental health well into their sixties, seventies and eighties.

No two persons age in the same way or at the same rate, nor does biological age keep pace with chronological. Some people are old in their thirties and forties and others young in their seventies and eighties. Genes play a large role and strong influences are psychological health, family support and stimulating work and hobbies. Some organs wear out faster than others, especially if there is an inherited or congenital weakness, or damage by abuse, disease or accident.

As the years roll by, the body is continually repairing itself and signs of aging are the gradual decreasing of its efficiency to do this. We just don't know how to control this, although nutrition and emotional attitudes are vital.

Emotional needs don't change as people grow older. Modern society imposes pressures due to neglect of the elderly, and the attitude of younger people often isolates them. They need to be useful and appreciated. Work is often denied them, and the young don't realize that their elders still need love and affection.

Some oldsters, however, annoy younger people by their attitude. They consider that their years endow them with greater wisdom and the right to respect despite their own lack of respect for others. Moreover, aging encourages in some characteristics such as irascibility, pettiness, ego centricity and paranoid tendencies, which are exasperating and irritating.

There are cultures that revere older people, but over the past seventy-five years western society has increasingly ignored this tradition. Today, once the arbitrary sixty-five is reached, it is assumed that capabilities vanish. It is a matter of record that most people can continue on effectively for many more years, but unfortunately the younger generations do not understand this.

What causes senile dementia? Doctors and scientists differ on the cause, but now agree that it is a disease and not the normal aging process as was once the consensus. They traced Alzheimer's disease back in families for eight generations and found that simple Alzheimer's, the most common form, appears not to be hereditary. So there are other factors involved, perhaps environmental plus a genetic predisposition and perhaps a viral infection in rare cases. In addition there may be a deficiency of essential trace minerals or the accumulation of harmful heavy metals in the body. Any toxic state due to drugs or alcohol may increase the likelihood of dementia.

There are chronic conditions, like diabetes or uncontrolled high blood pressure, that can bring on earlier senility, so can diseases of the kidneys and heart, lung disorders or a condition that decreases the blood supply to the brain. Another is hardening of the arteries, arteriosclerosis, which is discussed in the following chapter.

Abuse of the body over a period of years contributes substantially to premature aging. Smoking, alcoholism and drug abuse, obesity, poor nutrition and lack of exercise interfere with normal functioning of the organs and systems that maintain homeostasis. The more this is upset by abuse, the harder the body has to work, and when it is upset, the body is vulnerable to the development of the above diseases that are precursors of senile dementia. However, contrary to public belief, scientific studies show that a life of hard work, filled with problems, frustrations and traumatic events, does not contribute to premature aging or the development of degenerative diseases or senile psychoses.

Senile dementia is not the only cause of the deterioration of intellectual function. Many elderly people rely on drugs to relieve their aches and pains, and if they go to several doctors and take prescribed drugs from each, this can result in many pills being taken by one person. In addition, some patients mix over the counter drugs with prescription drugs, or drugs prescribed for other family members are shared among them. Those prescribed for a younger person may be too strong for the older, resulting in serious side effects. Therefore, adverse drug interactions can occur and be hazardous without the doctors knowledge or control. Equally dangerous is the failure to take necessary medication.

Reactions to drugs are intensified by changes in the aged body's ability to handle them, and the interaction of some can produce symptoms of senile psychosis, which may be mistaken for the onset of Alzheimer's. Therefore, it is best to report to the doctor all the drugs that are being taken and let him decide which to eliminate and which to take.

Can Alzheimer's disease be treated? No, however, deterioration of the mind should never be considered a natural consequence of aging. First, a thorough medical, neurologic and psychiatric work-up is in order, because many brain diseases are treatable and reversible. Physical therapy, electric shock therapy, and medications to lift the mood, calm anxieties, lessen hallucinations and delusions are often restorative, although useless as a cure for Alzheimer's.

There is exciting news, however. Research reveals that a deficiency of the neurotransmitter called acetylcholine is seen as the possible cause of Alzheimer's. Known in its popular form as lecithin, it is abundant in soy beans, corn and egg yolks, and is available as a dietary supplement. Research on Parkinsonism, a neurological disease that produces shaky hands, slowness and stiffness of the muscles, reveals that the neurons that release dopamine, another neurotransmitter, have degenerated, causing a marked deficiency. On the theory that insufficient dopamine causes the disease, how can it be replaced or the neurons stimulated to produce more? L-Dopa, the precursor of dopamine, is a substance that goes into the production of dopamine and patients given L-Dopa orally are greatly relieved of their symptoms.

Researchers find that the neurons that use acetylcholine as a transmitter have decreased by seventy to ninety percent in Alzheimer's patients, showing a deficiency or an inability to produce it, just as there is a deficiency of dopamine in Parkinson's patients. They are now working with acetylcholine on the same premise, that a dietary supplement could benefit the patient. So far this is only theory, but further studies support it.

Now the search is on to identify the substance that can be administered to the patient orally, intravenously or by injection, like insulin is to diabetics, so the neurons that produce acetylcholine in the brain cannot degenerate and fail to supply this substance. Furthermore, it must pass through the blood brain barrier in sufficient amounts to do its job. Therefore, the medium that carries it must be inactive so as not to harm brain tissue.

Researchers are investigating three avenues of approach. One, the addition of large quantities of choline to the diet, resulting in more choline in the neurons. This has been done, using choline or lecithin, which is choline in its natural form. Measurements of choline in the blood and spinal fluid proved this method feasible. Two, the administration of the drug physostigmine, which splits acetylcholine, thereby prolonging and strengthening its action. Three, the administration of drugs known to activate the receptors that are responsive to acetylcholine and thereby make the brain more responsive to the limited amount available.

So far, patients show little improvement with these three methods, and what does occur is very subtle. A bright note, however, is marked restoration of memory and some reversal of symptoms in patients treated with physostigmine in the early stages of the disease.

There is a drug, tacrine, developed in Europe, which awaits approval by the Food and Drug Administration. It seems to be partially effective in relieving the severity of the symptoms, however, studies show that it causes damage to the liver after long use.

Not long ago, pregnenolone, a hormone which is made naturally by the human body, has been found to greatly improve memory. For many years, it has been used as a steroid and found to be effective and safe, but was relegated to the back burner because it was not as effective as cortisone and prednisone. Furthermore, it was not considered a memory enhancer until recently when a scientist began studying its effect on laboratory animals and discovered that it has remarkable memory-enhancing powers. Pregnenolone will soon be tested on Alzheimer's patients and healthy elderly people who are having difficulty remembering and results of this study should be available in a year. The hormone appears to have a profound effect on brain cells and, as all hormones decline with age, supplying them may help to reverse the symptoms of aging.

Thus, intensive research is in progress to find the cause, prevention and treatment of Alzheimer's disease, particularly because of the high incidence in the population. Research determines without a doubt that families with early onset share an abnormal gene on chromosome 21 and a mutation on this gene directs cells to produce amyloid, a crucial nerve cell protein. This was found in the brains of Alzheimer's patients at autopsy, suggesting that this altered amyloid protein was responsible for the deterioration of brain cells. Other researchers have found that a gene on chromosome 19 is also implicated in patients from different families who have inherited this gene, which points to another possible etiology of the disease.

This does not mean that every member of the family will develop Alzheimer's, only those who carry one of the defective genes. It is interesting that a similar genetic defect is also found in Down's syndrome patients, and one gene is the same as a certain protein that forms plaques in the brains of Alzheimer's and elderly Down's syndrome patients. Abnormal deposits of this same protein is found in the brains of a variety of aged mammals, so not only human beings acquire Alzheimer's, animals do too. Dogs, polar bears, monkeys, rabbits and other mammals carry the risk, so work on animals greatly facilitates research.

In every case of the disease, autopsy demonstrates that amyloid fragments attach themselves to brain cells causing them to die, so research is concentrating on finding a drug that can prevent amyloid from attacking them. Another approach currently under study is to treat the glial or support cells in the brain. These are also destroyed in Alzheimer's disease, and this destruction has been reversed when treated with nerve growth factor.

Research in this area is new and much more must be learned. At least we know what the disease is, what is needed to keep it from progressing, and hopefully how to prevent it in high risk cases. In the final analysis to date, this combined evidence leads one to hypothesize that genetics, infectious agents and the environment, alone or in concert, is the probable cause of Alzheimer's dementia.

HOW DO YOU COPE WITH AN ALZHEIMER'S PATIENT?

At first, you don't realize her condition is serious, but slowly and insidiously little signs appear that are almost imperceptible. You are accustomed to have a drink together before dinner, and lately she is irritable and annoyed at subjects you usually talk about. This is also true when she is tired or under emotional pressure. She has always been well organized, but now she is forgetful, and this confuses her, for it is foreign to her nature, and she cannot understand why she is compulsive about things she normally pays no attention to. One day, while driving to the supermarket, she couldn't recall where it was.

You don't dismiss such changes in behavior, no matter how slow and subtle. Something is happening and you want to know. After a thorough physical and psychiatric examination and several tests, the doctor explains that your wife has beginning Alzheimer's disease, and what to expect.

At this stage, a patient is not difficult to care for, and the doctor recommends drugs to slow the progress. Tranquilizers control occasional attacks of violence, and the anti-confusional, neuroleptic and antiparanoid drugs help in selected cases. Rarely are antidepressants of benefit, and all drugs should be carefully monitored so the patient doesn't become oversedated and develop physical illness. As time goes on, however, it may be necessary to have help in the home, or you may have to place her in a nursing home. This you should be prepared to do.

As the illness worsens, her symptoms become exaggerated. She is more forgetful, particularly of recent events, and more confused. She pays little attention to grooming and is untidy and careless about her dress and cleanliness. She makes a big fuss over little things and pays little attention to important matters, such as health and finances. She neglects the present and recent past and focuses more and more on long past events, living her life over.

She is unable to take care of herself in matters of safety, may insist on driving the car or going out for a walk alone and crossing streets without looking. She must be watched, as she is careless and can get hit by a car or otherwise injured.

As time goes on, your wife becomes intolerant of all except members of the household. She cannot deal with outsiders, doesn't remember their names or recognize them. She focuses entirely on issues of no consequence which is annoying to all. After a while, she doesn't recognize family members.

She may be incontinent and pay no attention to it. She will demand some unimportant service, such as having her hair brushed. She is like a naughty child, refuses to go to bed, or gets up in the middle of the night and wanders in and out of the house. She may resist the help you offer, and when feeding her, throw the food in your face. However, fits of violence are rare but unpredictable. Most of the time she is lethargic.

Coping with this is enormously difficult, but compassion and understanding is required of all who come in contact with her. Caring for her can create friction. She may get up in the wee hours of the night and demand attention, refuse to eat when the meal is served, object to everything that is expected of her and refuse medications.

As with all patients whose minds are affected by disease, the formula is the combination of love, compassion and firmness, with understanding mixed in. So concoct a formula and stick to it. She cannot help the way she is, so must not be blamed or criticized. However, some actions need not be tolerated and this must be made clear. For the most part, she will accept this, but must be constantly reminded that the family works together as a team and she is a part of the team. Her memory grows poorer day by day, and when she argues and refuses to conform to the routine of the household, be understanding, but firm and consistent. It is not easy, and is a constant pressure on everyone.

Here are suggestions on how to relieve this pressure. While memory loss is mild, list the patient's daily schedule with time and place. Lay out clothes for the next day. Put pills for medication in a divided box labeled when to take them. An alarm clock can serve to jog the memory. Changes in environment are disorienting, so it is best not to move the patient from the accustomed room or house.

Patients often don't remember simple duties. Place signs, such as FLUSH over the toilet, TURN OFF over the faucets, DON'T TOUCH over the stove, etc. Patients may forget how to get to the bathroom. Take him or her there periodically, and make sure the diet and fluid intake are adequate to ensure good elimination.

Alzheimer's patients feel isolated, frustrated, angry and despairing. Frequent affectionate physical contact offers security and an alleviation of these feelings.

A patient may not be able to read, but still may be able to play a musical instrument, sing, dance or listen to music. Encourage this.

Patients should wear an I.D. bracelet in case he or she wanders from home, and neighbors should be told of the patient's condition.

How do you handle financial matters? Your wife is unpredictable with money and may spend it foolishly on something she doesn't need and balk at buying necessities. If she has access to considerable amounts, and the illness encroaches on legal or financial matters that may lead to trouble, consult a lawyer and ask whether power of attorney or a conservatorship should be established. She will object strenuously, however, and put up a fight against efforts to take this responsibility out of her hands. Let the physicians and lawyer make this decision, but don't wait too long. Oftentimes she is lucid and signs checks and gives orders in a rational manner, only to retract them the next moment and tear up the checks. When confronted, she is confused and irrational, won't pay the nurses, the doctor or the hospital and the family is caught in the middle.

So, a very vital part of protection of the patient and the family is to ensure that this doesn't happen. How do you do that? The most effective technique is to establish a conservatorship. A conservatorship is a legal procedure, and the laws differ slightly in each state. Consult the patient's or your attorney and be guided by his advice. And don't wait too long, for it may be months before a conservatorship can be established and a conservator appointed by the Probate Court to be in charge of the affairs of a person who is no longer competent to do so. This is necessary when a person loses mental faculties because of a disease such as a stroke, Alzheimer's, brain tumor, or brain injury.

There are two courses to take in forming a conservatorship, voluntary and involuntary. The voluntary is when the person asks for help because he feels unable to manage his affairs, and asks the court to appoint someone as conservator. This is an easier route than the involuntary, which is usually initiated by a family member who realizes that the individual is no longer competent. The next step is to obtain the doctor's affidavit of the person's condition and to petition the Probate Court to appoint a guardian to investigate the case. This is usually a lawyer, who stands between the petitioner and the patient, talks to him and his doctors, and presents a report to the court indicating whether or not he feels it appropriate for a conservatorship to be established.

If the court decides to do so, a lawyer is appointed to review the situation from the standpoint of the patient, and asks the patient if he would mind having someone looking after his affairs. This person is usually someone he knows. As a rule he agrees, as often he is not aware of what is going on and has no strong feelings one way or the other and is relieved that someone is concerned.

When there are no relatives, or no one interested or living near enough to serve as conservator, who petitions the court? It may be a social agency, a neighbor, the physician or a friend. Sometimes relatives can be traced, or a neighbor notices that the person is acting strangely. In one case, the neighbor called the department of health and a social worker was sent to the house. The patient had no relatives, so the bank was appointed as conservator and an officer assigned to the position.

There are two types of conservators, for the estate and for the person. The one for the estate is responsible for handling the assets and financial affairs, making investment decisions, paying bills. The one for the person is charged with making day to day decisions about the patient's well-being and care, what services should be provided, and to work with the doctor on the medical care. If the patient should be considered for nursing home placement, a conservator always has the final say and can override a senile patient's objections.

Does the conservator get paid? Yes, unless the fee is waived which is usually done in the case of a family member, who can serve both for the person and the estate. The standard fee is vague and described as "reasonable". When it is a bank or institution, the fee is based on the fiduciary service schedule, like that of a trust for managing assets and providing custodial services.

To whom is the conservator responsible? The Probate Court has supervision over the conservator, who must render accountings every three to five years. When the conservatorship is terminated by recovery or death of the patient, he has to file a final accounting once he is discharged from his duties, and if the court or others suspect mismanagement of the estate, the conservator is responsible, but this has to be proven.

In cases where there are few family members or the family is disinterested, it is common for a nurse or an aide to become close to the patient, treat him or her with kindness and somehow persuade the patient to turn over his or her estate. These patients are subject to undue influence and seldom know their best interests. Here is a typical case.

It became evident to her lawyer that Joan Havens, an elderly widow, was declining during the past five years. She became more and more disoriented and kept getting into difficulties that required legal intervention. These ranged from minor motor vehicle accidents to involvement with fly by night contractors with whom she signed agreements they did not fulfill. She seemed less and less able to cope with problems and forgot appointments and other responsibilities, until her finances and health were in jeopardy.

This progressed to the point that neighbors called the social service department and a case worker reported to the lawyer that there was a problem. The lawyer called her two daughters, a doctor was consulted and it was decided that a conservator should be appointed. Neither of the daughters was able to take charge and it was resolved that the bank should serve. The patient did not object, in fact she was much relieved, and the Probate Court appointed a guardian. The daughters placed her in a nursing home where she remained until she died, receiving the proper care and attention that allowed her to preserve her dignity and self-respect, as well as to save her estate from the unscrupulous vultures who prey on the elderly.

23 ARTERIOSCLEROTIC PSYCHOSIS

This is another form of senile dementia. It is a psychosis, and the technical term is cerebral arteriosclerotic dementia. "Cerebral" means of the brain, "arterio" of the artery, and "sclerosis" means hardening. So cerebral arterioclerosis is hardening of the arteries of the brain and is literally a thickening of the lining of the arteries which affects their elasticity, resulting in a narrowing of the passage through which the blood flows, diminishing adequate supply to the brain. This condition occurs in the smaller arteries.

A similar condition called atherosclerosis, from the Greek "ather" or chaff, is characterized by patchy, fatty plaques of matter that cling to the lining of the arteries, narrowing the passageway and making the heart work harder to pump the blood through. This type affects the larger blood vessels, and is the precipitating cause of heart attacks.

When these conditions appear in the brain, psychoses follow. We will consider them as one in this discussion, since the resultant brain disorder is the same for both.

No age group is completely free of changes in the blood vessels due to arterio- or atherosclerosis. It is seen in children and young males. Soldiers in their early twenties who die in battle have arteriosclerotic arteries in surprisingly large numbers. However, the majority of cases appear from fifty-five to seventy years of age, the next largest number after seventy. As paradoxically as it shows up in the young, it often fails to appear in the very old.

How does it start? The first signs are headache, dizziness and discomfort in the head and neck. These are followed by transient numbness and weakness in the arms and legs, difficulty in speech, trembling and short, jerky steps, and intolerance to alcohol. It is obvious that blood flow is being restricted to essential organs like the heart, brain and kidneys. Arteriosclerosis in the body does not necessarily mean that it is also in the brain, nor is it always present in the body if it is in the brain. The two conditions seem to be independent.

The disease usually leads to stroke from hemorrhage, caused by the rupture of a blood vessel in the brain, or the blockage of an artery by a thrombosis, which is made of the fatty plaques that build up in the arteries. This stage may begin with a series of tiny strokes or a more severe one resulting in partial paralysis.

Is the mind affected? Yes. Irritability and outbursts of weeping or laughter may precede a stroke several weeks before it occurs. After a stroke, symptoms appear suddenly or slowly, and differ in individuals. Some patients experience extreme fatigue at mental work, loss of creativity and impairment of memory. It may be difficult to carry out the simplest tasks. They have episodes of forgetfulness, irritability, depression and confusion, succeeded by periods of what appear to be normal, but at a slightly reduced level. This can go on for many years, and often starts as early as the fifties or sixties. Symptoms are milder in the mornings and worsen as the day passes. As the disease progresses, patients become confused, bewildered and agitated. Those with a paranoid makeup are fearful, suspicious and hostile, and may be aggressive.

Concern about a relative's ability to handle his personal affairs can trigger an appeal to the court for power of attorney. Patients have been known to be perfectly lucid and to perform brilliantly in court in the morning. Under emotional pressure, their minds seem to clear temporarily, only to relapse by afternoon into incoherence, irritability, agitation and delusions that someone is persecuting them.

Inexorably, the disease progresses and changes in personality take place. The patient neglects his appearance and is disinterested in other people and his surroundings. Personality traits, such as aggression, paranoia and hostility become exaggerated and uncontrollable. Many suffer from emotional instability which evolves into severe anxiety, and hallucinations cloud their minds and affect their behavior.

Suicide is uppermost in many minds, so patients should be watched carefully.

The severity of a patient's symptoms is variable and depends largely on personality makeup and the extent of the disease. One study shows that of a hundred hospitalized patients, forty-nine were sufficiently improved to return home. There are mild cases that last many years without progressing, but the majority are severe and, on the average, die within three months to two years after hospitalization. However, remissions have been known.

One kind of arteriosclerotic brain disease is multiple stroke dementia. It takes the form of minor strokes from tiny hemorrhages or blood clots which don't as a rule cause major psychiatric disturbances. The damage is mostly to the motor area, resulting in a limited physical disability. However, patients become difficult, complaining, whining and hypochondriacal. The extent depends on personality makeup.

A few patients have vague and non-specific psychiatric effects. They are mildly depressed and experience vague dissatisfaction, a wandering hopelessness and a hyponchondriacal need for treatment. For this, they run from one doctor to another.

Generalized arteriosclerosis in the brain usually results in primitive behavior, such as sexual, excretory and so forth, in other words, a lowering of civilized values. Patients eat sloppily and may lose control of the bladder or bowels. They are unaware of the feelings of others and less sensitive to their needs. Men are bemused by sexual thoughts and touch their granddaughters or feel young women in public, the "dirty old man" syndrome. A precise, polite, fastidious person turns into a slob, careless, dirty, unfeeling and unresponsive.

A slow and gradual change in personality takes place, with a general flattening of feelings, loss of sharpness, clear thinking, and primitive responses. You observe this when you visit nursing homes. As you pass the patients in wheel chairs, one reaches out to touch your hand, but another grabs and pinches it viciously, a primitive need to hurt, the next acts babyish and does mean little things All this reflects the personality. As the intellectual capacity diminishes the mind regresses and the patient becomes less and less a whole person.

Can these patients be helped? Yes, but this type of dementia requires psychotherapy and treatment by drugs. Knowledge of the personality is helpful in determining the right treatment to prescribe so the patient can live as comfortably as possible. Medical treatment is important to detect physical disorders and changes in life-style are recommended, such as good nutrition, cleanliness, exercise, adequate fluid intake and elimination. Depression can be alleviated with antidepressant drugs and individual or group therapy.

What are the causes of this disease? Uncontrolled high blood pressure (hypertension) is associated with arteriosclerosis, and a high level of cholesterol in the blood. It is believed that heredity also plays a role, as studies show a predisposition to hypertension and high blood cholesterol in families.

We don't know all the answers, it may be partly genetic, partly the result of improper diet, the lack of exercise, overindulgence in tobacco and alcohol, obesity or a combination. Many overweight people live long and healthy lives, but others, who abuse their bodies for thirty or forty years, develop diseases like diabetes, heart disease and hardening of the arteries. Then they run to the doctor demanding to be made healthy. What can be done about this?

Most of us are born with good health and to maintain it is a lifelong process, and many people remain healthy into their seventies, eighties and even nineties. How can this be done? Maintain your good health throughout life in order to prevent the development of disease. How can we maintain health and prevent disease?

1. Have a physical examination every year and follow your doctor's recommendations.
2. Choose a low fat diet with lots of fresh fruits and vegetables, carbohydrates, such as whole wheat bread and pasta, brown rice, potatoes, and low fat dairy products. Keep the sweets at a minimum.
3. Don't smoke.

4. Don't overindulge in alcoholic beverages and never use hard drugs.

5. Exercise at least three times a week and keep your weight close to what it was in your twenties.

HOW DO YOU COPE WITH A PERSON WITH DEMENTIA?

It is difficult to adjust to living with a person like this and hard to accept that the mate you knew for so long is no longer the same. You keep feeling that this will go away and he will be his old self. Try to understand that his mental function is reduced, his coordination also, so he cannot think or move as fast or as efficiently, and this causes frustration and his personality and his intelligence to change.

So don't ask as much of him or be angry when things go wrong. Be compassionate but don't baby him or give in to him on every score. Be firm and make sure he understands what is expected of him, and that he is to behave in an acceptable manner. Treat him as you would a child and be cross if he doesn't do the right thing. However, always be aware that there is a gradual lessening of capacity in all areas. However, you don't have to put up with the manipulative, hostile and unreasonable behavior that occurs at times. Just turn your back and walk away, and don't try to settle old scores or take revenge for past misdeeds.

Gentle discipline is effective, as many of these patients can reason. Their awareness is good, and they can still learn.

24 STROKE

What is stroke? A stroke is when a section of the brain is being deprived of blood. Blood carries oxygen, which is essential to the brain cells and to life itself.

A stroke can take one of three forms, the most common is when an artery to the brain is blocked by fatty plaques. This is called atherosclerosis or hardening of the arteries, and is similar to a heart attack when arteries in the body are partially or totally closed, preventing blood flow to the heart muscle. In addition, blood clots (emboli), tend to develop in an artery filled with this fatty material, because blood forms clots when it collides with a foreign substance. And these clots lodge in the artery and slow down or stop the flow of blood.

The second form of stroke is called cerebral (brain) embolism, which is a blood clot or embolus that wanders to the brain from somewhere in the body and blocks an artery. The result is the same as with atherosclerosis, deprivation of blood carrying oxygen to the area of the brain fed by that artery.

The third form, cerebral hemorrhage, is when an artery in the brain bursts, flooding the tissue around it. Thus, the area involved is damaged, and interferes with the workings of the brain.

How the patient is affected by a stroke depends on the location of the damage and how widespread it is. A stroke can be slight or severe or somewhere in between. It can cause paralysis of one side of the body, loss of speech, one of the senses or the ability to understand the spoken word. It may affect behavior, memory or how one thinks.

What causes a stroke? High blood pressure (hypertension) is the most common cause when it is not kept under control by drugs. Pre-existing heart disease and diabetes can also bring on a stroke. Proper diagnosis and treatment of these conditions offers the best chance of preventing one and most can be prevented.

Are there other risk factors? Yes, and they are important. You may not have a condition described above, most people don't. However, we have long known that smoking constricts the blood vessels, and can lead to stroke if combined with another risk factor, such as overweight, a high intake of cholesterol and fatty foods and lack of exercise.

What should I do to avoid having a stroke? Have your blood pressure checked regularly and don't smoke. Eat a diet low in cholesterol and fat, and high in foods such as chicken, fish, turkey, beans, low fat dairy products and lots of fruits and vegetables. And exercise. Exercise is important because it helps to keep your weight down and stimulates the circulation. This strengthens the heart and helps to carry off waste products that tend to clog arteries. All the above is preventive medicine and prevention is the best way to avoid stroke.

So stroke can be prevented, and this has been proven in the last two decades. The death rate from stroke has fallen forty-five percent since 1970. How did this happen? Partly because of new tests and treatments, and because Americans are moving into healthier life-styles such as described above. However, stroke still remains high as a cause of death among the older group and accounts for many nursing home admissions because they were not conversant with preventive measures.

What is the incidence in the population? After heart disease and cancer, stroke is the third leading cause of death in the United States. It is the main cause of disability. About half a million Americans suffer a stroke each year, of which 145,000 die and the same number are severely disabled. If you have one stroke, you are at risk of another.

How do I go about this change in life-style? First, eliminate the risk factors. Again, what are they? High blood pressure (hypertension), hardening of the arteries (arteriosclerosis), heart disease, diabetes, smoking and overweight. The most important is high blood pressure. This can be controlled by medication and a change in life-style. Often a mild or borderline case can be eliminated simply by a change in life-style. If you are overweight, reduce to a level recommended by your doctor, by eating less, eating the right foods and exercising as well.

Studies show that eliminating salt and adding supplementary calcium to the diet may bring a mild case of high blood pressure down to normal levels. Recent studies eliminate salt as a causative factor of high blood pressure. Time will finally nail down the real culprits.

What else can I do? If you smoke, stop. Smoking constricts blood vessels, and puts an added load on your heart, making it difficult for the blood to flow easily and efficiently throughout your body.

Arteriosclerosis comes from overweight and over indulgence in high cholesterol foods, such as beef and pork, bacon, pastries, fried foods, lots of cheese, butter and heavy cream. And most importantly the lack of exercise of the aerobic type.

If you have diabetes or heart disease, see your doctor regularly and follow his orders so these conditions are treated and controlled.

Develop a lifelong habit of regular exercise. This is tremendously important not only to prevent stroke but many conditions that we are subject to, especially in the older years. Our bodies were built to move, and the technological revolution has made it possible for man to move less and less. So he puts on a few pounds every year and keeps them. The secret is to eat a little less each year, and exercise in order to maintain your optimal lean weight.

Are there warning signs of a stroke? Yes, and it may be a TIA. What is that? TIA stands for "transient ischemic attack", and is an explicit warning that a stroke may occur. How do you know you have had a TIA? There may be numbness or weakness in an arm or leg, it may be difficult to talk, there can be unusual headaches, dizziness or trouble in making decisions. These symptoms are temporary and disappear after a few hours. What causes this? There is a slight interruption or reduction of the flow of blood to the brain. It is imperative to report symptoms such as these promptly to your doctor.

What can be done once a stroke occurs? Rehabilitation is the answer, and is successful in restoring the ability to function in most patients. The extent of the success is determined by the amount of damage that has been done to the brain, how large an area was affected, the attitude of the patient, and whether the patient thinks positively or negatively. Other important factors are the competence, skill and expertise of the rehabilitation team and the cooperation of the family. With an A plus situation, a patient can eventually experience reversal of symptoms to a large extent, often completely.

What does rehabilitation consist of? Rehabilitation is started when the patient is in the hospital and includes physical therapy to strengthen muscles. Speech therapy, if this is impaired, is instituted and occupational therapy to restore the ability to care for oneself. The process takes from a few weeks to months and can extend into years, depending on the individual and the extent of the brain damage. Many patients recover completely. At least two years of recovery should be vigorously attempted before accepting disability.

In addition to rehabilitation, new tools are being discovered in the treatment of stroke patients. Research has recently developed a substance to inject into the spinal fluid that reverses the results of the stroke, if applied within the first few hours. This holds great hope for stroke patients, but the best approach is prevention.

HOW DO YOU COPE WITH A STROKE PATIENT?

Although it is difficult to care for this person, it is important that you understand what is happening to his personality and to be aware of the frustration he experiences when the ability to take care of himself is suddenly taken away.

Love is the most important ingredient in his recovery, and the more you can give, the more he improves. Rehabilitation is important also, but loving care influences recuperation more than any other element. This is true particularly during the first two to three years, which is when the maximum repair takes place. The brain can relearn what it partially lost and progress may continue for many more years. The wonders of medical science are also helpful, but the one factor that matters most is the affection and supportive loving care of a close relative or friend.

But don't neglect yourself, for it is also important to understand that you too have needs, apart from the person you love. You may feel resentful about the demands that are made on you and your patient can be very childish. He finds fault easily and is critical of how you take care of him. Often he is angry, at what you don't know, and protests that he is being abused.

His physical abilities are limited and when you try to help, he is explosive and emotional, making it difficult for you to manage. In handling this you need tolerance and firmness, as with a child. You should have time to yourself, or you may wind up angry and resent the illness in a way that may be unfair. At times you are fed up, and cause an emotional collision that you later regret, for this upsets the patient a great deal. You have been together a long time and love one another, and this is very hard to take.

What is happening to his brain? When there is sufficient injury, the patient reverts intellectually to a primitive level, the extent determined by his basic personality, i.e., whether he is more of a thinker or a doer. At this level, his responses are more animal like. For example, he may refuse to bathe or brush his teeth or have someone do it for him even though he can. There is lack of comprehension. This has gone completely and it makes you

angry because you are only trying to help.
What do you do with your anger? Do you hit him or shout or swear? Or do you keep it inside and pretend you accept everything he says and does? So what should you do? For when you swallow anger, what happens? You get depressed. The best is to work it off in a physical way, exercise or see your friends and family, go to a movie, try to find enjoyment and release. A change of pace will do it.

Some people feel taking care of a loved one is enough, and don't take care of themselves or their own emotional needs. This is not wise, for you may be doing more than smothering anger. You may have financial needs, interpersonal reactions, sexual needs, and you may be isolating yourself to the detriment of your well-being.

How you react to this depends largely on the relationship between you and your spouse over the years. Who was dominant? Who was the most aggressive? Which one found pleasure for him or herself more? Who was the babied one? Who was the cared for one? When an illness reverses these rules, a relationship may come into conflict. Then therapy is useful for the healthy spouse to make certain he or she is not smothering feelings or overreacting to them.

Is the quality of life important to the patient? Some say it is, and if this is lost don't want to go on living. Others want to live and be cared for as long as necessary. Some are not aware of what has happened, and survive with all bodily systems functioning well, but their brains have ceased. This applies not only to stroke patients but to all who are brain damaged by senility, Alzheimer's disease, injury, tumor or arteriosclerosis.

It takes a strong mind to say: "Life is worth nothing, and lie down and die. Many can do this, many cannot. Those who cannot are unaware that they are a bother to all concerned, and cling desperately to life because that is all that is left. Then the primitive instinct begins to work, they breathe, the hearts beat and the organs function, but the intellectual qualities disappear, and they have no influence over what they say or do. They are totally out of contact.

VI UNUSUAL AND RARE DISORDERS OF THE BRAIN

PREAMBLE

Serious study of the anatomy of the brain began in the early part of the 17th century with the philosopher René Descartes. His work marked a turning point in the understanding of the nervous system. He separated thinking from the anatomy of the brain, thus dividing the mind from the body. When phrenology became popular in the 19th century, it attempted to rejoin the mind and body which Descartes had separated. Subsequently, the science of neurology studied the location of certain functions of the brain such as language, the senses, emotional states and motor skills.

Descartes' concept remained in vogue until this day when it has become recognized that the mind and the body interact and influence one another. We now know that psychological and/or physical stress can reduce immunological responses. In other words, both can lower the body's resistance to disease.

As the 20th century appeared, two different opinions developed. One was the "localization" theory whereby certain functions are performed in specific areas of the brain and the other was the "global" view that the brain works as a whole, even though specific areas of the brain are responsible for certain functions.

The improvement of the microscope brought big advances. Cells were discovered and staining made it possible for scientists to speculate that messages go from neuron to neuron through electrical impulses. Further research produced electro-physiological techniques through microelectrodes, by which electrical currents could trace impulses from cell to cell. Thus, it was discovered that several neurons work together to transmit particular impulses.

Next came the realization that not only was there electrical activity of the neurons but also complex chemical phenomena, and the production of neurotransmitters which cross synapses and give information to the next cell. Then, in the 1950s it was found that too many or too few neurotransmitters cause disturbances and changes in the emotions, and a morphine molecule and other molecular groups can treat pathological states such as anxiety and depression.

Next came the formulation of theories on memory. It was suggested that when a neuron sends a message to the dendrites of another neuron, the second neuron becomes sensitized to the message. These neurons are part of a ring that becomes a reverberating circuit of an electrical current running in a circular motion. This creates short-term memory and, if they are repeated over and over, permanent changes in the protein structure of the neuron is made, resulting in long-term memory. So memory is situated in the synapses, and since brains have billions of synapses they amass enormous data banks.

Recent researchers question Sigmund Freud's concept of memory that information is stored in the unconscious to be brought up at will. Today it is maintained that there is no such place as the unconscious or any fixed storage place for memory. They postulate that no memory is ever the same but the brain constantly recreates and changes it, and combines information in an endless way. Every memory has something in common but when recalled it may not be the same as it was originally and is perhaps colored by emotion.

Freud's theories became classical neuroscience but are now being replaced by new concepts. The brain seems to reinvent data each time it is faced with using its memory and what is remembered is never quite the same. So the unconscious is not controlling our minds. Each time we think of a memory it may come out a little different and that is why there are distortions of thought. We recreate the thought the way we want it.

25 MULTIPLE PERSONALITY DISORDER

This fascinating and exotic condition was first described by Morton Prince, M.D. eighty years ago and its description has not changed since. Due to its complexity, however, and its relatively uncommon occurrence, it is often misdiagnosed and it remains controversial whether this phenomenon really exists. Furthermore, an overwhelming majority of psychiatrists have never identified a case, while a few encounter them fairly often. The American Psychiatric Association's "bible", the Diagnostic and Statistical Manual, offers three and a half pages on multiple personality disorder (MPD) in the latest edition of 1987, whereas in earlier additions it was merely mentioned and never even alluded to before 1980.

This differential attests to several possibilities, one, that MPD is now more accurately diagnosed, two, that it ceases to be confused with other mental illnesses such as schizophrenia, three, there is an increase in the incidence because there is more child abuse, and four, it has become a fad of psychiatry.

One group estimates the incidence in the population to be one percent of the psychiatric cases, and more women than men are afflicted by a ratio of from three or more to one. Others have another opinion as to the incidence and hold that it is a rare disorder. So the controversy rages on.

Just what is multiple personality disorder? The essential feature is that there exists within one individual two or more distinct personalities, some even of the opposite sex. All the personalities are not total, but some are what is termed personality states, which means that they are not fully developed. There are always two or more of the personalities that control the life of the individual and are fully developed, and the states wait in abeyance to take over in a special situation, such as an emotionally or environmentally induced instance. The average number of fully developed personalities ranges from eight to thirteen within one person and the number of states can be up to a hundred.

Personalities switch from one to another within seconds to minutes, often triggered by psychosocial or situational stress. Some may be aware that there are other personalities that do not communicate with them, while others interact normally, and some may be unaware that there are others. At any given time, only one personality interacts with the outside world.

Interestingly, two different personalities may represent opposite personality characteristics, such as, one may be a seductive, flamboyant temptress, while another a Victorian type, shy spinster, and the two may alternate, much to the confusion of one another. Other opposite traits show up, such as passive and aggressive, and a cruel, paranoid person versus a caring, comforting one. There is no limit to the diversity of these personalities, as there is no limit to the combination of personality traits in normal human beings.

The personality that is "out" most of the time is called the "host" personality and is not always aware of the existence of others. This results in mental confusion because of gaps and conflicts in daily life. Nevertheless, most MPDs lead fairly normal lives and marry, have children and maintain careers as well as complete high academic degrees.

An extraordinary phenomenon of the multiple is the changes in abilities and attributes. The personalities in one multiple may be left-handed, another right. One may weigh more than another and have different talents, such as artistic ability, another at the piano. One may speak German, another will not understand it and can speak only English.

Another interesting feature of the multiple is that one personality can complain of a physical illness and another may not show any sign of it. For example, one personality has allergies and another does not, one said she burned her arm and the burn was evident, another said she had been beaten and there were lash marks on her body. When the host switched to another personality, there was no evidence of burns or bruises. One person had high blood pressure, another did not. When she switched to a third personality, she suffered severe headache and the symptoms of hypertension subsided in a few minutes. A woman was on insulin for diabetes and required different amounts of insulin depending on the personality that was in control.

One multiple was anorexic and when out, starved herself. The other personalities would come out and eat well. Changes in sight were also observed requiring modification of eyeglass prescriptions for each personality.

The personalities sometimes form twosomes, and stick together according to age, such as adolescence or early childhood. They have different names, both given and last names, at times with symbolic meaning, such as "Melody", the personality expressing itself through music. Some have no name, or simply the name of its function, such as "The Protector".

What causes this fascinating disorder? Several studies indicate that in almost all cases of multiple personality disorder there was abuse in childhood, often sexual, or some form of severe emotional trauma. The incidence seems to be increasing, in 1979 there were 500 cases reported, in 1990 five to six thousand have been identified. Why? The assumption is the apparent increase in child abuse. Fortunately, not every abused child develop MPD. When queried, ninety-five percent had experienced abuse and ninety-two percent had considered or attempted suicide.

What is the treatment for multiple personality disorder? Once properly diagnosed, the therapy is psychotherapy. It is a lengthy and grueling ordeal for both therapist and patient because multiples lost faith in people when they were very young and continue to be distrustful of opening up to anyone, especially an authority figure. So, therapy is painful and intensive, three times a week for three or more years. It consists of delving into the original hurt which the patient had succeeded in burying deep within the psyche, and reliving those events is devastating. Hypnosis is helpful and encourages the different personalities to share the pain. When this occurs, the personalities that are most similar come together and are no longer separated. However, this does not come easily, for when it does, all of a sudden there appears another personality not heretofore encountered, and the process has to begin all over again in order to fuse this last one to the others.

Notwithstanding these hurdles to surmount, the prognosis is encouraging, although there are few studies and the groups are small. One therapist, who has had much experience treating MPDs reports a success of ninety percent out of fifty-two patients. Success was based on two years of having only one personality and showing no sign of any more.

Here are a few case histories. Julia went for help when she was thirty years old, the usual age most multiples go for advice. She was concerned about her lapses in memory and time over the years, during which presumably one of her alter egos took over and acted in her stead. In Julia's words: "I remember when I was in the third grade and went back to school after Christmas break, and all of a sudden it was October and I was in the fifth grade." Now, in her late twenties, she remembers being in a bar, a place she never goes alone, talking to a stranger. The conversation led her to believe the man thought she was there for a pick-up. Horrified, she looked down at herself and realized that she was dressed like a prostitute.

Julia became preoccupied about her breaks in reality and sought help. Now she knows that she has several alter egos who can take over and that she has control of the switches from one to the other. She is a happier person.

A psychiatrist treated Sybil for ten months for depression. Every so often, she appeared for a session with her wrists bound up, and on examination they were found to have been slashed. When asked how this happened, she said "I don't know", and went on to explain that funny things were going on she didn't understand. One day she found clothes in her closet that she would never have bought or wear. She was a school teacher and these dresses were scanty and revealing, and she didn't know how they got there. Furthermore, there were cigarette ashes in her car and she doesn't smoke. The doctor recalls that she came to his office one day dressed like a streetwalker with a cigarette hanging out of her mouth. It was at this session that the true diagnosis of multiple personality was made.

Multiples differ in the reaction of their personalities to medications, foods and environmental influences, as demonstrated by Robert, who had an alter ego named Tommy. Tommy was allergic to citric acid and when Robert drank orange or grapefruit juice, he had no allergic reaction. If, however, he went "in" soon after drinking it, Tommy and the other personalities would break out with itching and develop blisters. When Robert returned, the itching disappeared, although the blisters remained.

This is Brenda's story. "I am forty years old, was married when I was nineteen and have three daughters. I have fifty-three personalities, called alters, some of whom are children. Once in a while, one of the children comes out and I am the mother. Another one is Tina, who is not a nice person. She is seventeen, opinionated and at times can be vulgar and rather wild. She has a nasty disposition and likes to come out and have fun.

When I was married it was to one of my personalities, a young child, not the real me. My husband now tells me that way back then I had temper tantrums and loved to color so, at my insistence, we bought coloring books and crayons. He liked to color too, so we got along well.

I had no idea what was the matter with me for many years and at times thought I was crazy, and other times that everyone was like me. I have always heard the voices in my head and thought that was normal and that everyone did too. But I couldn't explain the things that went on in my life. When I was older I began to realize that all this was not normal at all. So I went for help. Before that I was afraid to go to a doctor for fear he might put me away.

The therapy opened up my mind and I remembered that I was abused sexually when I was three years old by both my father and my mother. It was very painful to recall. Sometimes when the doctor asks who raped me, I answered: "I didn't get raped, Sally did." Sally is one of my young personalities.

This is Beatrice's story. She was severely abused as a child but learned to cope with her condition and to control it. Her personalities don't come out automatically since she mastered how to control them and to summon them up when and if she wants to. When she brings out her child personality named "Princess", her face and voice change completely and take on that of a child.

Both her parents had multiple personality disorder and she remembers as she was growing up that she could go along with whatever personality they would switch into. However, she has not spoken to them in many years and never wants to again. She doesn't believe that they love her as they abused her so terribly, so she doesn't love them any more, now that she knows the truth.

These are two women who have achieved commendable success in life regardless of having multiple personalities. Kathleen is a real estate manager in a large city and was diagnosed six years ago with one hundred and sixty-five personalities. She is under therapy and has never shared her condition with anyone but her doctor. Jacqueline is another. She is a lawyer in a large city and has a successful practice. No one knows of her multiple personalities. She has a hundred, and they include a man, a gay, a straight and a very young person. Her father is a multiple also and both have been abused as children.

Obviously, both of these women have a chance of resolving their disorder if they stay with therapy, as ninety percent of cases have reached resolution permanently.

26 DISSOCIATIVE DISORDERS

What does this mean? To dissociate means to sever oneself from association with a person or activity. In relation to a mental disorder, it means to be removed and apart from the person's self or his awareness of reality.

Here is an example. A woman fell from a third story balcony and fractured her pelvis. After the accident she related her perception of what happened. "I experienced standing on another balcony watching a pink cloud floating down to the ground. I felt no pain at all and tried to walk back upstairs." She had a remarkable sense of remoteness and separation from the actual event.

The psychiatric manual treats dissociative experiences as rare, but recent studies strongly suggest that these bizarre phenomena are common in about five to ten percent of the population. The symptoms are feelings of unreality, of being an automaton. Some people suffer amnesia, which is partial or total loss of memory. This places dissociative disorders in the category of depression and anxiety, but as these emotions are often fleeting, subtle and elusive, they are seldom reported to the doctor. Furthermore, it is believed that if reported they are often misdiagnosed.

Our brains are extraordinary creations and it is interesting to note that we all experience mild symptoms from time to time of dissociation. Studies find that eighty-three percent of people realize that they have not heard part of a conversation and twenty-nine percent occasionally feel as though they are watching themselves in a movie. About fourteen percent of those surveyed look into the mirror and don't recognize themselves. Only occasionally do any of these sensations occur, but a few say this happens about one-third of the time. Thus, many normal people have mild symptoms of dissociation and attach no importance to them.

The disorder appears to be arranged on a spectrum, with the mild symptoms at one end, visiting many individuals, and the severe symptoms afflicting only those with psychiatric problems. They are all elusive, subtle, silent and fleeting, and involve a strange absence of feeling or curious sense of unreality. About forty percent of patients who suffer from depression, anxiety, panic attacks or schizophrenia acquire this disorder.

These experiences are a common reaction to life-threatening events, such as an automobile accident, rape, intense trauma of all kinds or physical abuse. The most common symptom is depersonalization, which is a sensation of being separated from your body. You feel as if you are somewhere else, like a robot that has no feelings. It is an unreal experience. Sometimes it is like watching a movie of yourself going through life.

Once in a while normal people experience these emotions. A study of college students finds that thirty percent reported occasional episodes. They were not under the influence of alcohol or drugs when they were queried, but it is more common when they are toxic. Two-thirds of the victims of the San Francisco earthquake of 1989, a group of ninety men and women, experienced mild symptoms of dissociation. So it appears that these sensations are normal responses to intense trauma. A very mild but annoying example is the common occurrence of remembering something that you want to do in another room, going there and forgetting what you went there for. By going back to where you thought of it, you remember. This happens to all of us and is annoying but of no consequence. It is merely human.

27 THE IDIOT SAVANT OR THE SAVANT SYNDROME

The dictionary states that a "savant" is a man of profound or extensive learning, so the term "idiot savant" is paradoxical. However, that is exactly what an idiot savant is. He has a low I.Q., yet one portion of his brain is highly developed. He has extraordinary capabilities for extremely intricate and difficult performances which it requires a person with a high I.Q. and high energy to attain after years of study and practice. The savant is already there with no study or practice.

The word savant comes from the French "savoir", meaning "to know". The word idiot is the cut off level of the I.Q. test but most of these patients have an I.Q. of 40 ranging up to 55 and are usually in the lower range. They also vary in the ability to demonstrate the precocity and extraordinary extent of the talent with which they are endowed, whether it be music, mathematics, drawing or other talent.

Along with low intelligence, savants have one or two outstanding skills such as calendar calculation, the arts and music. They are mostly males and occur with high frequency among the autistic population. Those who have the most impressive calendar calculating skills have a strong interest in calendars and devote considerable periods of time to dates and their relationship to other dates. Patients with this gift are often socially withdrawn.

The savant syndrome was first described in 1886 by Dr. Down, after whom Down's syndrome is named which is a form of mental retardation. Since then, many cases of savant syndrome have been reported. They are categorized into two groups, the prodigious and the talented. The first definitive writing on this fascinating subject was in 1914 by Tredgold. His careful, detailed descriptions remain unsurpassed and have been duplicated over and over, however, since the syndrome was first described a century ago, the combination of severe mental handicap and phenomenal mental ability is still a mystery.

Here are highlights of a few of Tredgold's patients. There are six males to one female. Many are born blind or with another sensory handicap. All are extremely mentally retarded except for one or two areas in which they have an exceptional and unprecedented capability. mostly in music, mathematics or the visual arts. Some savants can calculate four figures with lightning rapidity or sing a song with perfect words and pitch after hearing it once. One deaf savant has outstanding skills in drawing, invention and mechanical dexterity. Who taught them? How did they learn? Extensive research cannot offer an iota of why or how.

The ratio of six males to one female, the same as in autism, suggests that the disorder may be sex-linked. There are many theories to explain this but not one suffices so far. They fall into categories such as photographic memory (eidetic imagery), but this one is not evident in all savants, as it also shows up in persons whose brains have been damaged when the imagery persists beyond adulthood.

How often in the population does this fascinating syndrome occur? One fact we are sure of is that it is very rare. A survey done in 1975 of twenty-three facilities with a total population of 90,000 mental retardates shows that it occurs in about one in two thousand. Although autism is known to be linked to the savant syndrome in isolated cases, there is no substantive information on this available. Here are some examples.

The extraordinary memory for data recorded on calendars which we have already mentioned is fairly common in savants. For example, an imbecile boy of twelve can multiply three figures by three figures in half a minute. Another area is music and a crippled, blind girl can repeat perfectly the words and tune of any song played or sung for her once. In the arts, a severely mentally deficient boy can draw cats so skillfully and beautifully that he is dubbed "the cat's Rafael". One of his pictures was bought by King George IV of England. Another case was that of a patient in an asylum who possessed extraordinary skills in drawing, inventions and mechanical dexterity.

Lightning calculating and mathematical ability are possessed by many savants, including a blind imbecile who could solve algebraic problems with no knowledge of geometry. Many similar cases have been recorded. One cretin could repeat verbatim the articles in a newspaper he had read and another could repeat the same backwards. There were twins who were calendar calculators who could remember the weather for every day since they were eighteen. Some savants have artistic talents, and a few have become famous, such as one previously mentioned and one in Japan, called the "Van Gogh of Japan".

Musical skills are also prevalent in savants. These patients seem to have a combination of blindness, musical genius and mental retardation. One, Tom, had a low I.Q., but could play Mozart at the age of four without a mistake and any piece after he heard it once. Also, he could repeat a conversation in any language without the loss of a syllable.

Another case is that of a female musical savant who, at the age of six months, was humming operatic arias. The savant with musical aptitude seems to be limited to playing the piano, singing or humming with perfect pitch and recall. Often skills occur in combination. Music and memory are the most commonly seen.

A thirty year old sculptor with remarkable talent achieved considerable acclaim in art circles. His I.Q. is 40 and his subjects are life-sized animals. First he worked in clay then proceeded to bronze. A handicapped boy has remarkable musical talent which began when he was three. His vision is impaired and he can speak only in simple sentences. He has had no piano training but his repertoire is huge, and he can perform both modern and extremely complex classical pieces with perfect pitch and accuracy and can improvise after a single hearing.

A thirty-six year old blind man who is also hemiplegic is a highly skilled pianist with a large repertoire and an impressive knowledge of music. He has had no formal training but can transpose, improvise and imitate. In life he functions at a level of severe mental retardation.

How do savants do these remarkable things? There are many theories and several speculations that have been expounded over the years but none can be proved to be universal. Autopsies of the brains of savants show abnormalities in certain areas but none of this is conclusive. There is a theory that heredity plays a part and that the savant has inherited two factors, one for retardation and the other for an extraordinary skill or talent. Various studies conclude that a special talent may run in the family but this factor is not seen in all savants. It may, however, be a contributing cause but not the explanation. So the search continues.

Unique and at times prodigious memory combined with severe mental defects, coupled with a talent and/or skill, is the hallmark of most savants. They have superior memory and it is interesting to note that paranoid schizophrenics also have highly developed and unusually acute memories.

Modern sophisticated techniques are opening the door to learning more about the savant than can be observed by autopsy and leading the way to unraveling this fascinating enigma. With high tech diagnostic tools such as MRI, CAT scan and PET, scientists are now able to look closely at the brain. These tests show prenatal influences on the left hemisphere, one of which is a male related factor, perhaps testosterone. This slows growth and allows enlargement of the right brain which harbors the skills of the savant, namely mathematics, music and visual arts. This could explain the right brain skills and the predominance of males who are savants.

As far as we know today, the answer to this enigma lies in the disruption of the functions of the brain's left hemisphere to the extent that it does not or cannot fulfill its normal role. This disruption may result from prenatal influences or injury and could be sex-linked, leading to a migration of neurons to the right hemisphere in order to compensate, causing the right to dominate the left. So the abnormality of the left hemisphere is coupled with circuitry probably arising from the same influences, and thus the savants' capabilities, defects and extraordinary memory are produced.

Genetic factors must be involved as well in prodigious savants, for how could the access to the vast knowledge and rules of music, art and mathematics, which seem to be innate, be accounted for? There is also an intricately intertwined connection between autism and the savant, although all autistics are not savants or the other way around. Some abnormalities of the brain appear in both subjects, but a separation point is not clear.

The savant's memory is without consciousness and is very deep and very narrow, whereas our memory is broad and wide and filled with emotion. His is limitless and fixed but with no emotion, of which he is incapable. This demonstrates an entirely different circuitry within the brain. Our memory is exceedingly wide and heavily laden with emotion. It is more limited than that of the savant but flexible and able to associate with other experiences. It can also be creative; the savant's memory cannot create.

New technology, brain mapping and new knowledge about the brain, gives us hope that we will be able to find the key and unlock the secrets of the savant, his prodigious memory and fascinating abilities and this knowledge will lead us to knowing ourselves better.

We admit that a genetic factor may be the etiology, partially or wholly, for the savant syndrome, but we also entertain the thought that his unprecedented capabilities may have come from a former life. Could he have been a genius or a talented individual in music, mathematics or the visual arts?

INDEX

A

Acupuncture, 192
Addiction, 168-196
Aging, 210
Agoraphobia, 61
Alcoholics Anonymous, 179, 181, 192
Amphhetamines, 173
Alcoholism, 178-181, 183-196
Anger, 30, 50
Alzheimer's disease, 71, 73, 75, 208-217
Anorexia nervosa, 126-131
Anxiety, 46-48, 61-64
Arteriosclerotic psychosis, 208, 211, 218-221
Attention deficit, 137-138
Autism, 133-136, 233-234, 236
Autonomic functions, 16-17

B

Barbituates, 173
BEAM See Brain electrical activity mapping
Behavioral learning, 16
Biological psychiatry, 19-21, 26
Brain, 22, 28-30, 32, 35, 41, 44, 48, 50, 52, 65, 69, 71-77, 92, 96-97, 105, 112, 114-115, 124-127, 129-130, 134, 142, 148, 151, 155-158, 160, 163, 165-168, 180, 183-184, 188, 196-197, 202-203, 208, 212-219, 221, 224, 226-227, 235-236
Brain electrical activity mapping, 96
Brain imaging, 71-75, 92, 96, 124-125, 196, 199, 235
Brain injury, 190
Brain tumors, 196
Bulimia nervosa, 128-131

C
Caffeine, 176-178
CAT scan See Computed axial tomography
Chemical therapy, 19, 26-27
Childhood psychoses, 132-138
Churchill, Winston, 38, 118, 210
Claustrophobia, 61
Cocaine, 35, 77, 168, 171-173, 178, 181-184
Cognitive psychology, 14
Compulsive perfectionism, 32
Computed axial tomography, 92, 196, 235
Computerized tomography, 72-74, 96, 199
Congenital abnormalities, 136-137
Conservatorship, 215-217
Crime and criminals, 12-14, 17-18, 32-35, 144-157, 182
CT See Computerized tomography

D
Davanloo, Habib, 25
Delusional paranoid psychosis, 96-103
Depression, 12, 49-51, 58, 62, 73, 104, 112-121, 123, 127-129, 147, 190
Descartes, René, 226
Dissociative disorders, 231-233
Doing [personality trait], 11-12, 28-33, 37-40, 45, 55, 57, 91, 97, 106, 132,
    139, 145, 152-153, 165, 170, 180, 187, 198, 200, 202-203, 209, 224
Down's syndrome, 78, 136, 205-206, 213, 233
Dreams and dreaming, 29
Drugs, 34-35, 73, 77, 151-152, 167-184

E
Ego, 19
Energy, 29, 41-43, 50, 76, 113-115
Environment, influence of, 13, 15-16, 150, 186, 205
Epilepsy, 197-201
Exercise, 30
Exhibitionism, 162-163

F
Fragile X syndrome, 206
Freud, Sigmund, 12, 14-15, 17, 19-21, 227
Frotteurism, 163

G
Genetic engineering, 77-79
Group therapy, 23-24, 26, 60

H
Habit, 14-15, 123
Hallucinogens, 173-174
Head injury, 201-204
Heredity, 15, 18-19, 45, 85-86, 138, 183, 185, 205, 220, 235-236
Hitler, Adolf, 38, 147
Homeostasis, 35, 45, 65-66
Homosexuality, 157
Homosexuals, 158, 161-162
Hughes, Howard, 124
Hyperactivity, 137-138
Hyperventilation, 59, 61
Hypnosis, 56, 229
Hypochondria, 51-54
Hysteria, 54-56, 203
Hysterics, 54

I
Id, 19
Idiot savant, 136, 233-236
Insecurity, 48-49
Instinct, 15-17, 19
Intelligence, 29, 40-41

J
Jung, Carl, 15

K
Korsakoff's psychosis, 195, 202
Kraepelin, Emil, 71, 79

M
Magnetic resonance imaging, 72-74, 96, 196, 199, 235
Manic-depressive, 11, 71, 103-111, 190-191
Marijuana, 174-176
Memory, 14-15, 17, 123, 227, 235-236
Mental illness, 11, 20-21, 28, 69-131, 200
Mental retardation, 204-208
Mesmer, Franz Anton, 19
Metabolism, 75, 104
Mood, 30, 49-51, 75-76, 104, 110-112, 120, 143, 146, 166
Movement disorders, 137-138
MRI See Magnetic resonance imaging
Multiple personality disorder, 227-231
Munchausen's syndrome, 154-155

N
Neuroses, 11-12, 44-64, 73, 203-204
Neurotic tendencies, 44-45

O
Obsessive compulsiveness, 44, 56-58, 73, 120-126
Opium, 173
Oral compulsion, 171
Panic attacks, 60-64, 73
Paranoia, 29, 36-38, 45, 47, 51-54, 58, 80, 96-103, 128, 145-148, 155, 202, 219
Parkinsonism, 212
Passive-aggressive, 29, 44, 187
Pederasty, 162
Pedophilia, 161-162
Personality, 28-44
Personality disorders, 139-164, 227-233
PET See Positron emission tomography
Phobias, 58-60

Positron emission tomography, 71, 73, 75, 96, 124-125, 199, 235
Prince, Morton, 227
Psychiatry, 12, 19-21
Psychoanalysis, 12-14, 19-20, 24-27
Psychopaths, 33-34, 69-72, 75-76, 132-138
Psychoses, 11-12, 69-72, 75-76, 132-138
Psychosomatic illness, 54, 64-68
Psychotherapy, 21-27, 48, 50-51, 54, 56-58, 86, 88, 115, 117, 127, 129, 134-135, 142, 191-192, 229

R
Rape, 159-161
Regional cerebral blood flow, 73
Rush, Benjamin, 19

S
Savant syndrome See Idiot savant
Schizoaffective disorders, 111-112
Schizophrenia, 11, 71, 73, 79-96, 111, 191, 227, 235
Schumann, Robert, 105
Seasonal affective disorder, 118-119
Senile dementia See Alzheimer's disease; Arteriosclerotic psychoses
Sex offenders, 155-164
Sexual harassment, 161
Sexuality, 12, 19-21, 29, 85, 158
Short-term dynamic psychotherapy, 24-27
Single photon emission computed tomography, 73, 75
Somatopsychic illness, 64, 66, 68
SPECT See Single photon emission computed tomography
Stroke, 204, 218-225
Superego, 19
Sydenham, Thomas, 139

T
Thinking [personality trait], 11, 28-32, 37-40, 45, 57-58, 79-80, 88, 97, 106, 132, 139, 152, 165, 170, 180, 187, 198, 200, 202-203, 209, 224
Tics, 137
Tissue plasminogen activator, 78

Titchener, Edward Bradford, 14
Tranquilizers, 177-178
Transient ischemic attack, 223

U
Unconscious, 13-19

V
Van Gogh, Vincent, 80
Voyeurism, 163

W
Watson, John B., 14
Wundt, Wilhelm, 14

Z
Zoophilia, 163

www.ingramcontent.com/pod-product-compliance
Lightning Source LLC
Chambersburg PA
CBHW020050170426
43199CB00009B/237